Introduction to
Industrial Controls
and Manufacturing

Academic Press Series In Engineering

Series Editor
J.David Irwin
Auburn University

This series includes state-of-the-art handbooks, textbooks, and professional reference books in engineering. A particular emphasis of the series is given to the applications of cutting-edge research in industry. The aim of the series is to bring together interdependent topics in electrical engineering, mechanical engineering, computer engineering and manufacturing which are essential for success in modern industry. Engineers, researchers and students will find these books to be a necessary part of their design toolkit.

Published books in the series:

Control in Robotics and Automation: Sensor Based Integration, 1999, editors B.K. Ghosh, N. Xi and T.J. Tarn
Industrial Controls and Manufacturing, 1999, E. Kamen
DSP Integrated Circuits, 1999, L. Wanhammar
Time Domain Electromagnetics, 1999, S.M. Rao
Single- and Multi-chip Microcontroller Interfacing for the Motorola 68HC12, 1999, G.J. Lipovski

Introduction to
Industrial Controls
and Manufacturing

Edward W. Kamen
School of Electrical and Computer Engineering
Georgia Institute of Technology
Atlanta, Georgia

ACADEMIC PRESS
San Diego • San Francisco • New York • Boston
London • Sydney • Toronto

This book is printed on acid-free paper ∞

ACADEMIC PRESS
525 B. Street 1900, San Diego, California 92101-4495, USA
http://www.apnet.com

Academic Press
24-28 Oval Road, London NW1 7DX, UK
http://www.hbuk.co.uk/ap/

Library of Congress Cataloging-in-Publication Data

Kamen, Edward W.
    Industrial controls and manufacturing/Edward W. Kamen.
        p. cm. — (Academic Press series in engineering)
    ISBN 0-12-394850-9
    1. Automatic control. 2. Process control. 3. Manufacturing processes
    — Automation.   I.   Title.   II.   Series.
TJ213.K25   1999
670.42'75 — dc21   98-52664
                                                         CIP

Printed in the United States of America

99 00 01 02 03   MB 9 8 7 6 5 4 3 2 1

To my granddaughter Erin

Contents

Preface

1   Introduction   **1**
Manufacturing Fundamentals   1
Introduction to Control   3
Problems   6

2   Example of a Process Control Problem   **9**
Continuous-Variable Control   12
The Laplace Transform   15
Return to Level Control   16
Effect of Output Flow   21
Problems   22

3   Modeling of Continuous-Variable Processes   **27**
Transfer Function Representation   28
System Modeling Based on the Step Response   31
Input Delay   38
Problems   41

4   Control of Continuous-Variable Processes   **47**
Closed-Loop Configuration   47
Tracking a Step Reference   50
PI Controller   51
PID Controller   59
Effect of a Disturbance   62
Processes with Input Delay   66
Problems   69

5   Digital Control   **75**
The z Transform   77
Design of Digital Controllers   80
An Alternate Design Approach   87
Problems   90

6   Model Predictive, Adaptive, and Neural Net Controllers   95
     Model Predictive Controllers   95
     Direct Adaptive Control   104
     Neural Net Controllers   106
     Problems   110

7   Discrete Logic Control   113
     State Diagram Representation   114
     Boolean Logic Equations   121
     Generation of Boolean Logic Equations   128
     Problems   134

8   Ladder Logic Diagrams and PLC Implementation   141
     Electrical Ladder Logic Diagrams   142
     Software Ladder Logic Diagrams   147
     PLC Implementation   151
     Bottle-Filling Operation   156
     Problems   162

9   Manufacturing Systems   165
     Performance Measures   166
     Flow-Line Analysis   171
     Flow-Line Analysis with Machine Breakdowns   178
     Line Balancing   181
     Problems   186

10   Production Control   193
     Production Control Schemes   194
     Pull Systems   196
     A Push-and-Pull System   201
     Problems   203

11 Equipment Interfacing and Communications   205
     Equipment Interfacing   205
     Communications   209
     Web-Based Studies   214

Appendix A Further Reading   217

Appendix B Laboratory Project   219
     Process Demonstrator   219
     Project Description   221

Preface

Industry applications often involve continuous-variable process control and discrete logic control in a manufacturing environment. Although a number of textbooks exist on traditional (continuous-variable) control which is taught in engineering schools, there is a lack of treatments on control that combine continuous-variable control, discrete logic control, and manufacturing fundamentals. In an attempt to fill the void, this book contains an introductory treatment of the essential topics including analog and digital control, discrete logic control, ladder logic diagrams, manufacturing systems, and production control. The material in this book is based on an undergraduate engineering course that was developed and taught by the author at the Georgia Institute of Technology. A description of the course was presented at the 1997 American Control Conference in Albuquerque, New Mexico, and was published in the proceedings of the conference.

There is enough material in the text for a three-credit-hour quarter or semester course. A semester course version may require supplemental material if three 1-hour lectures are given per week. In a semester system, the best format is to have a three-credit-hour course with two 1-hour lectures per week and a 1-credit-hour laboratory project. A detailed description of a laboratory project is given in Appendix B. The project provides students with hands-on experience in using programmable logic controllers (PLCs), PC-based controllers, and software for equipment interfacing, operation, and communications. The book is intended to be appropriate for junior- or senior-level engineering students or for practicing engineers. The background for reading the text consists of some previous exposure to calculus, Boolean algebra, and the concepts of signals and systems. The part of the book dealing with continuous-variable control does involve the use of the Laplace and z transforms which are introduced in the text. It is helpful if the reader has had some past experience using transforms, but the brief treatment on transforms in the text is intended to be sufficient for the application to control system analysis and design considered in the book. MATLAB is used in the text to generate plots, compute step responses, etc., but no previous experience with MATLAB is required.

The book begins with an introduction to manufacturing and control in Chapters 1 and 2, and then goes into continuous-variable control in Chapters 3 through 6. A key feature of the continuous-variable part is a development of a modified PI controller that allows for the assignment of the closed-loop zero due to the controller. This result, which is known but appears to be a well-kept secret, provides a powerful method for achieving a desired transient performance when tracking a reference signal. In Chapter 6 a brief introduction to advanced control techniques is given, including model predictive control, adaptive control, and neural net control. Discrete logic control and PLC implementations are considered in Chapters 7 and 8. A systematic procedure for designing discrete logic controllers is presented in terms of a state diagram for each of the state variables describing the desired control action. Boolean logic equations are generated from the state diagrams and then the equations are implemented on a PLC using ladder logic diagrams.

Manufacturing systems and production control are considered in Chapters 9 and 10. Various performance measures for manufacturing systems are given in Chapter 9, and in Chapter 10, production control is characterized in terms of the concepts of push-and-pull systems. Analogies with standard open-loop and closed-loop process control are given. The last chapter of the book deals with equipment interfacing and communications with a brief introduction to OPC, the GEM standard, fieldbuses, and Ethernet. Appendix A contains a list of textbooks for further reading, and Appendix B contains a description of a laboratory project based on a process demonstrator. There are homework problems at the end of Chapters 1 through 10, and suggested web-based studies at the end of Chapter 11.

The author wishes to thank George Vachtsevanos for conceiving the idea of the process demonstrator and his efforts in overseeing the design and development of the process demonstrator at Georgia Tech. Thanks also go to Alex Goldstein for his efforts in the construction of the process demonstrator and his input on the two-tank system that is described in Appendix B. Special thanks go to my former student Mike Gazarik who developed and taught the project laboratory for the course. Thanks also go to Dan Creveling, who served as a lab instructor for the course; to Michael Barnes, who developed a tutorial for the lab; and to Payam Torab, who provided input on the material on discrete logic controllers. Finally, thanks go to Marc Bodson for pointing out the modified PI controller, Ken Cooper for his comments on the text, Chen Zhou for use of his notes on ladder logic diagrams, and Mike Vitali for attending student project presentations and offering his comments.

Edward W. Kamen

Chapter 1

Introduction

Manufacturing deals with the conversion of raw materials into finished materials or products. As illustrated in Figure 1.1, a manufacturing operation can be viewed as a manufacturing system with inputs equal to the raw materials and with outputs equal to the finished materials or products. The range of materials and products produced today is incredibly vast and expanding every day. Examples include metals, glass, plastics, and other materials, chemicals, textiles, pulp and paper products, food and beverages, home appliances, automobiles, and electronics such as computers, digital cameras, and communication devices. The types of processes that arise in manufacturing include machining such as drilling, cutting, and grinding, materials handling using conveyors and robotic loaders/unloaders, casting, painting or plating, parts assembly, molding, brewing or cooking, blending or mixing, and waste management.

Manufacturing Fundamentals

A large component of manufacturing deals with *discrete parts manufacturing*, where discrete parts are processed and assembled together to yield the final product. Examples include the production of automobiles, home appliances, and computers. An example of a discrete parts manufacturing system is the surface mount printed circuit board assembly line shown in Figure 1.2. Here the inputs to the system are printed circuit boards with metal wiring lines and metal pads for the location of circuit components. As seen from the figure, the manufacturing process consists of four steps: a stencil screen printer which deposits solder paste onto the pads; a placement machine which places circuit components onto

1

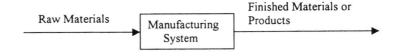

Figure 1.1   Input/output representation of a manufacturing system.

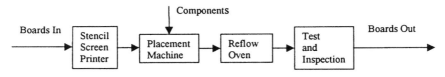

Figure 1.2   A surface mount printed circuit board assembly line.

the pad locations; a reflow oven, which melts the solder paste to form a permanent electrical and mechanical bond between the components and the pads, and a test and inspection station for checking the quality and performance of the completed board

Manufacturing also involves the processing of nondiscrete components such as liquid materials, chemicals, beverages, etc. This includes *continuous-flow manufacturing* in which the product is produced as a "continuous item." An example is the production of metal sheeting, which is illustrated in Figure 1.3. In this manufacturing system, metal ore is liquefied by a furnace, and then the liquid metal is allowed to cool and begin to take a thick flat shape. In the last two steps, the hot malleable metal is rolled into a strip having the desired thickness.

The four major components of any manufacturing operation are:

1. Business activities such as marketing and sales
2. Product design
3. Production planning
4. Production

These components used to be treated separately, but today there is a major emphasis on integrating them. For example, design is now based in part on production considerations; that is, designs are sought that yield the lowest possible cost of production. This is called *design for manufacturing* (DFM).

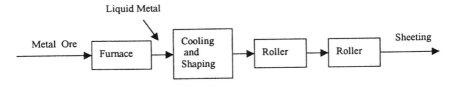

Figure 1.3   Continuous-flow production of metal sheeting.

All four of the components of manufacturing just listed are implemented using computer technology, resulting in *computer integrated manufacturing* (CIM). In Chapters 9 and 10, we cover some of the fundamental aspects of production systems. The topics of business activities, product design, and production planning are beyond the scope of this book and thus are not covered.

Production is achieved via *automation* that involves the use of control technology for operating devices, equipment, and processes in a manufacturing system. Although control technology arises in other application domains in addition to manufacturing, the major emphasis of this text is on the use of controls in a manufacturing environment. Since manufacturing is carried out using equipment and processes appearing in industry (i.e., in companies that manufacture materials or products), the controls that are considered are often referred to as "industrial controls." This is the reason for the use of the term in the title of the book. We conclude this chapter by giving a brief introduction to control.

Introduction To Control

The application of controls in manufacturing is expanding at a very rapid pace due to the desire to reduce production costs and to produce higher quality products that perform better. By "control," we mean automated control that is carried out by devices called *controllers*, as opposed to manual control that requires human intervention. Today, almost all (automated) controllers are implemented using microprocessors, digital signal processing (DSP) chips, programmable logic controllers (PLCs), personal computers (PCs), or workstations.

A key component of control technology is the design of the controller. Common types of controllers include proportional-plus-integral (PI), lead-lag, model predictive controllers (MPCs), adaptive controllers, neural net controllers, and fuzzy-logic controllers. However, controller design is actually a relatively small part of the overall control problem. Other issues that are of major importance in control technology are:

- Signal conditioning and signal processing such as filtering and sampling
- Use of real-time and historical databases to store information
- Interfaces between controllers and sensors, devices, and equipment
- Operator interfaces including graphics user interfaces (GUIs)
- Communication networks such as fieldbuses and local-area networks (LANs)
- Computer operating systems such as Windows NT
- Object-oriented (OO) control software such as Java-based applications

Interfacing and communication networks have become extremely important in control applications. An introduction to these topics is given in Chapter 11.

When applied to manufacturing, control is used on each of several different levels in the hierarchy shown in Figure 1.4. The top level is the enterprise level where high-level functions are carried out involving finances, administrative matters, customer relations, etc. Next is the factory floor level, which contains the complete production facility, and then the cell or line level that contains groups of machines. The bottom level of the hierarchy is the machine/process level, which contains the individual equipment items needed to manufacture the product.

We first consider control on the lowest level, the machine/process level, and then work our way up in the last part of text.

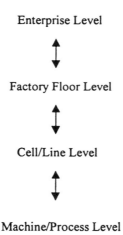

Figure 1.4 Control is used on each of these levels of a manufacturing system.

General Control Configuration

The general configuration for *closed-loop control* of a process (or system) is shown in Figure 1.5. Here $c_1, c_2, \ldots, c_m$ are the control inputs and $y_1, y_2, \ldots, y_p$ are the process outputs. The process has a total of m inputs and p outputs, where m and p are positive integers. In a typical application, m and p may be greater than 10 or even greater than 100. The configuration shown in Figure 1.5 is said to be of the closed-loop type since the process outputs y_i are fed back through the controller to generate the control inputs c_i to the process. The controller is usually realized using a microprocessor, DSP chip, PLC, or PC. The functionality of the controller is determined by a control recipe that can be downloaded to a PLC or PC before system operation. Reference signals may also be applied to the controller in order to specify desired operation.

The control inputs c_i and the process outputs y_i may be continuous variables or discrete variables. Here "continuous" means that the variables may take on values over some range of real numbers. Examples of continuous variables are voltages, position coordinates, pressures, flow rates, etc. Continuous variables are also said to be analog variables. "Discrete" means that the variable may have only a finite number of different values. An example of a discrete variable is the variable M, which represents the on/off status of a motor, and which is defined by $M = 0$ when the motor is off, and $M = 1$ when the motor is on.

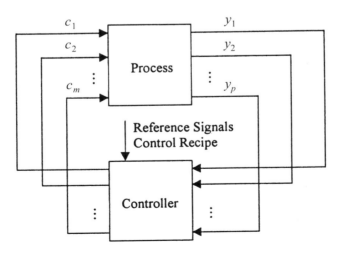

Figure 1.5 Closed-loop control configuration.

The control inputs c_i and the process outputs y_i may be functions of time t, in which case we write c_i and y_i as $c_i(t)$ and $y_i(t)$. When c_i and y_i are functions of t, the process or system is said to be *time driven*. This means that the system is driven by inputs that change values as a result of time t changing values. When $c_i(t)$ and $y_i(t)$ are continuous variables, the system is said to be a continuous-variable time-driven system.

Instead of being functions of time t, the c_i and y_i may change values due to the occurrence of events, in which case the system is said to be *event driven*. For example, suppose that the system is a motor that is controlled by a switch. The status of the motor is denoted by M, where $M = 1$ when the motor is on and $M = 0$ when the motor is off, and the status of the switch is denoted by S, where $S = 0$ when the switch is off and $S = 1$ when the switch is on. If we take the input to be S and the output to be M, the system (the motor and switch) is defined by the equation $M = S$, so that the motor is on if and only if the switch is on. This is an event-driven system since the input S depends on the event of turning the switch on or off. Since there is only a finite number of input and output values, this system is also referred to as a *discrete event system*.

A number of systems appearing in practice have multiple inputs and outputs, with some of them functions of t, and with the others event dependent. Such systems are said to be *hybrid systems*. Very little exists in the way of systematic techniques for analyzing or designing hybrid systems; however, a number of researchers are currently attempting to develop a theory for the study of hybrid systems.

Problems

1.1   Give an example of a discrete parts manufacturing system (not mentioned in this book) with at least three steps. For your example, describe the steps of the manufacturing process and give a block diagram showing the sequence of operations.

1.2   Give an example of a continuous-flow manufacturing system (not mentioned in this book) with at least three steps. For your example, describe the steps of the manufacturing process and give a block diagram showing the sequence of operations.

1.3   Give an example of a continuous-variable time-driven system (not mentioned in this book) with single input $c(t)$ and single output $y(t)$.

1.4 For your example in Problem 1.3, derive a relationship between the input $c(t)$ and the output $y(t)$. If a relationship cannot be generated, find another example (not mentioned in this book) where a relationship can be found and specify what it is.

1.5 Give an example (not mentioned in this book) of a discrete event system with two inputs C_1 and C_2 and one output Y.

1.6 For your example in Problem 1.5, derive a relationship between C_1, C_2 and Y. If a relationship cannot be generated, find another example (not mentioned in this book) where a relationship can be found and specify what it is.

1.7 A controller for an oven operates as follows: If the oven on-switch is activated, the oven door is closed, and the oven temperature is below a desired level, the controller turns on the oven heater. If the heater is on, or when the temperature is above the desired level and the door is closed, the controller turns on the oven fan. If the light switch is on or the door is open, the controller turns the oven light on. An oven-door limit switch indicates if the door is open or closed and a temperature limit switch indicates if the temperature is above or below the desired level.

(a) Define the inputs and outputs of the controller.
(b) Derive a relationship between the inputs and the outputs of the controller.

1.8 A gas furnace in a home is controlled by a thermostat. The temperature in the house at time t is denoted by $T(t)$. The status of the furnace is denoted by F, where $F = 1$ when the furnace is on and $F = 0$ when the furnace is off. The status of the thermostat is denoted by $THST$, where $THST = 1$ when the thermostat is closed and $THST = 0$ when the thermostat is open. When the thermostat is set at T_d degrees, where T_d is the desired temperature in the house, the thermostat closes when a decreasing $T(t)$ reaches the value $T_d - 1°$, and the thermostat opens when an increasing $T(t)$ reaches the value $T_d + 1°$.

(a) Describe how the furnace is controlled using the thermostat. Express the control operation in an equation form.
(b) The temperature $T(t)$ in the house can be modeled by the differential equation

$$\frac{dT(t)}{dt} = \alpha - \beta \text{ when the furnace is on}$$

$$\frac{dT(t)}{dt} = -\beta \text{ when the furnace is off}$$

Here α is the temperature increase in deg/min due to the heat provided by the furnace, and $-\beta$ is the temperature decrease in deg/min due to heat loss. The value of β depends on the outside air temperature, which we assume is fixed. The homeowner conducts an experiment on a typical winter day and finds that the temperature increases 0.5 deg/min when the furnace is on, and decreases 0.1 deg/min when the furnace is off. Determine the values of α and β in the model given above.

(c) Using the model with values found in part (b), derive an expression for the temperature $T(t)$ in the house for $0 < t \le 60$ min, assuming that $T(0) = 65°$ and $T_d = 70°$. Sketch $T(t)$ for $0 < t \le 60$ min.

(d) Does your result obtained in part (c) appear to be reasonable from the standpoint of what you would expect the temperature variation in the house to be? If so, explain why. If not, explain how the model could be modified to yield a better result.

Chapter 2

Example of a Process Control Problem

A common type of control problem arising in the process industries is the movement of fluids among tanks and the processing of the fluids in the tanks. An example of a two-tank system is shown in Figure 2.1. In this system, fluid can be pumped from Tank 1 to Tank 2, and then can be pumped back to Tank 1, and thus the system is "closed." In actual practice, the system would be "open," so that batches of the fluid being processed can be moved through the system. We are considering a closed system since it can be used as a demonstration testbed for control projects in a university or industrial laboratory setting. In Appendix B, we provide a detailed layout for a two-tank system along with a description of experiments that can be carried out on the system.

In the two-tank system shown in Figure 2.1, the status of the two pumps is denoted by $P1$ and $P2$, where $P1$ and $P2$ are discrete variables having the value 0 when the pump is off and the value 1 when the pump is on. The variables $P1$ and $P2$ are event driven since their values depend on the event of turning on or off the pump. The variables $iv(t)$ and $ov(t)$ are the positions at time t of the input and output valves for Tank 2, respectively. Both $iv(t)$ and $ov(t)$ are continuous time-dependent variables with values ranging from 0 to 1; that is, $0 \leq iv(t) \leq 1$ and $0 \leq ov(t) \leq 1$. The valves are closed when $iv(t)$ and $ov(t)$ are equal to 0, and they are completely open when $iv(t)$ and $ov(t)$ are equal to 1. The continuous time-dependent variable $f_{in}(t)$ is the flow rate in cubic inches per second of liquid into Tank 2, and $f_{out}(t)$ is the flow rate in cubic inches per second of liquid out of Tank 2. Both $f_{in}(t)$ and $f_{out}(t)$ are measured using flow-rate sensors. The continuous variable $y(t)$ is the level in inches of the liquid in Tank 2. The level $y(t)$ is measured using a pressure sensor located at the bottom of Tank 2 (see Appendix B).

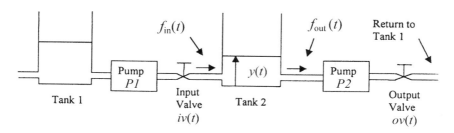

Figure 2.1 A two-tank system.

There is a constraint on the operation of the pumps; namely, the first pump is off ($P1 = 0$) whenever $iv(t) < 0.1$, and the second pump is off ($P2 = 0$) whenever $ov(t) < 0.1$. This constraint is imposed to ensure that the pumps are on only when there is fluid flowing so that they do not burn out.

Two examples of control problems that can be defined for the two-tank system are as follows:

Control Problem 1: Fill Tank 2 to a desired level L and keep the level at L even though the output flow rate $f_{out}(t)$ may be varying as a consequence of the output valve position $ov(t)$ being varied. This is a level control problem.

Control Problem 2: Fill Tank 2 to a desired level L_1, then heat the liquid to a desired temperature T, and keep the level at L_1 and the temperature at T for an interval I of time. Then after the time interval I, drain Tank 2 to level L_2, and repeat the steps over and over again with the smallest possible cycle time. This is an example of a *recipe control problem* where the recipe is the specified sequence of steps. The type of control required here is called *sequencing control*, which is usually implemented using a PLC. We pursue this type of control in Chapters 7 and 8.

We can solve Control Problem 1 using a discrete event controller defined as follows: First, let Y denote the discrete-valued variable defined by

$$Y = 0, \quad \text{if } y(t) < L$$

$$Y = 1, \quad \text{if } y(t) \geq L$$

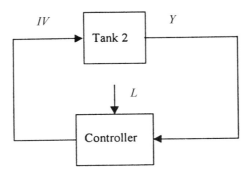

Figure 2.2 Discrete event level control of Tank 2.

By definition, Y is equal to 0 when the level $y(t)$ of the liquid in Tank 2 is less than the desired level L, and Y is equal to 1 when the level $y(t)$ of the liquid in Tank 2 is greater than or equal to the desired level L.

We can view the position of the input valve as a discrete variable with the value $IV = 0$ when the valve is closed and $IV = 1$ when the valve is open. Then in terms of the discrete variables Y and IV, we have the closed-loop control configuration shown in Figure 2.2. The controller in Figure 2.2 is given by the equations

$$\text{If } Y = 0, \quad \text{then } IV = 1$$

$$\text{If } Y = 1, \quad \text{then } IV = 0$$

By definition of the controller equations, we see that the input valve is open ($IV = 1$) when the level $y(t)$ in Tank 2 is less than the desired level L ($Y = 0$), and the input valve is closed ($IV = 0$) when $y(t)$ is greater than or equal to L ($Y = 1$).

Since the first pump cannot be on when $IV < 0.1$, this constraint needs to be included in the controller equations, which results in the equations

$$\text{If } Y = 0, \quad \text{then } IV = 1$$

$$\text{If } Y = 1, \quad \text{then } PI = 0 \quad \text{and} \quad IV = 0$$

The tank with this controller is an example of an event-driven system since the variable IV changes value as a result of the occurrence of the event that the input valve is turned on or off, and the variable Y changes value as a result of the occurrence of the event that the level $y(t)$ of liquid in Tank 2 exceeds the desired L, or L exceeds $y(t)$. Since the control is given by the above logic (if–then) equations with discrete values, the controller is often referred to as a *discrete logic controller*.

If this discrete logic controller were to be implemented (e.g., using a PLC), the level $y(t)$ of liquid in Tank 2 may vary substantially in a highly erratic manner due to the on/off switching of the input valve. This is due to the dynamics of the process and, in particular, to the time delay that exists between a control command applied to the input valve and the closing or opening of the valve. The time delay is considered later.

Continuous-Variable Control

Instead of using a discrete logic controller as discussed earlier, we can achieve level control using a continuous-variable controller. The first step to solving the level control problem using a continuous-variable controller is the generation of a model for the process, which is carried out next.

With r equal to the radius in feet of Tank 2, the cross-sectional area of the tank is equal to πr^2 square inches, and the volume $V(t)$ of liquid in the tank at time t is equal to $\pi r^2 y(t)$ cubic inches. The rate of change of the volume $V(t)$ is equal to the flow rate of liquid into the tank minus the flow rate of liquid out of the tank; or in mathematical terms,

$$\frac{dV(t)}{dt} = f_{in}(t) - f_{out}(t) \qquad (2.1)$$

Inserting $V(t) = \pi r^2 y(t)$ into (2.1) gives

$$\pi r^2 \frac{dy(t)}{dt} = f_{in}(t) - f_{out}(t)$$

and thus

$$\frac{dy(t)}{dt} = \frac{1}{\pi r^2} [f_{in}(t) - f_{out}(t)] \qquad (2.2)$$

Now the flow rates $f_{in}(t)$ and $f_{out}(t)$ are proportional to the valve positions $iv(t)$ and $ov(t)$, respectively, and thus we can write

$$\frac{1}{\pi r^2} f_{in}(t) = \alpha iv(t - t_d) \text{ and } \frac{1}{\pi r^2} f_{out}(t) = \beta ov(t) \tag{2.3}$$

where α and β are constants, given in inches per second, and t_d is the time delay in seconds in the response time of the input valve. The constants α and β and the time delay t_d can be determined from measurements performed on the two-tank system. Then inserting (2.3) into (2.2) results in the following differential equation model for Tank 2:

$$\frac{dy(t)}{dt} = \alpha iv(t - t_d) - \beta ov(t) \tag{2.4}$$

In the following analysis, we neglect the time delay t_d in (2.4), so that the differential equation model becomes

$$\frac{dy(t)}{dt} = \alpha iv(t) - \beta ov(t) \tag{2.5}$$

With Tank 2 given by the differential equation (2.5), the objective is to determine $iv(t)$ so that the level $y(t)$ of the liquid in the tank is equal to the desired level L. We want to design a continuous-variable controller that generates $iv(t)$ from measurements of $y(t)$ and the specification of L. The closed-loop system consisting of the tank and controller is shown in Figure 2.3.

In continuous-variable control, a more standard form for the closed-loop configuration is shown in Figure 2.4. In this setup the controller is in cascade with the process (the tank in this case), and the input to the controller is the error $e(t) = L - y(t)$ between the desired level L and the actual level $y(t)$. The signal $r(t)$ in Figure 2.4 is called the *reference*; in this case the reference $r(t)$ is equal to the desired level L for $t \geq 0$. The control configuration shown in Figure

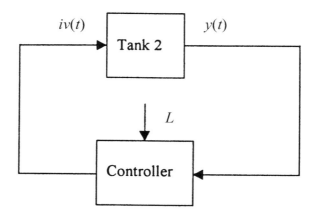

Figure 2.3   Continuous-variable level control of Tank 2.

2.4 is said to have *unity feedback* since the output $y(t)$ is fed back with unity gain to the differencer where $y(t)$ is subtracted from the reference $r(t) = L$.

The simplest type of controller is the *proportional controller*, or P controller, where the control signal [$iv(t)$ in this case] is proportional to the error $e(t)$ between the reference $r(t)$ and the output $y(t)$. In mathematical terms, we have

$$iv(t) = K_p e(t) = K_P [r(t) - y(t)] \qquad (2.6)$$

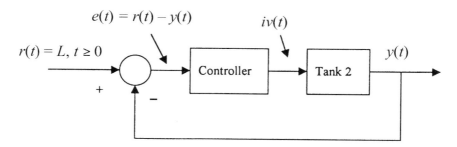

Figure 2.4   Standard form for continuous-variable feedback control.

where $r(t) = L$ for $t \geq 0$. The constant K_P is the gain of the proportional controller. Inserting (2.6) into (2.5) gives

$$\frac{dy(t)}{dt} = \alpha K_P[r(t) - y(t)] - \beta ov(t)$$

Rearranging terms gives the closed-loop differential equation

$$\frac{dy(t)}{dt} + \alpha K_P y(t) = \alpha K_P r(t) - \beta ov(t) \tag{2.7}$$

To simplify the analysis, we shall assume that the output valve is closed, so that $ov(t) = 0$ for all t. Then (2.7) becomes

$$\frac{dy(t)}{dt} + \alpha K_P y(t) = \alpha K_P r(t) \tag{2.8}$$

To determine how well the proportional controller works in this case, we need to find the solution $y(t)$ to the closed-loop differential equation (2.8). In particular, we would like to know if $y(t)$ will converge to the desired level L as t increases from $t = 0$, starting from initial condition $y(0) = 0$, so that the tank is empty at time $t = 0$. We would also like to know what the transient looks like; that is, how $y(t)$ increases from $y(0) = 0$ to its steady-state value assuming there is a steady-state value. To carry out this analysis, we need to use the Laplace transform, which is introduced below.

The Laplace Transform

Given a function $f(t)$ of time t, the Laplace transform (LT) of $f(t)$ is a function $F(s)$ of the complex variable $s = \sigma + j\omega$, where j is the square root of -1. We always denote the LT with an uppercase letter. The LT of a large class of functions can be computed from the LTs of a few "basic functions," such as the ones given next:

$$f(t) = c, \, t \geq 0 \rightarrow F(s) = \frac{c}{s}$$

$$f(t) = ct, \ t \ge 0 \to F(s) = \frac{c}{s^2}$$

$$f(t) = ce^{-at}, \ t \ge 0 \to F(s) = \frac{c}{s + a}$$

$$f(t) = b \cos \omega t \to F(s) = \frac{bs}{s^2 + \omega^2}$$

$$f(t) = b \sin \omega t \to F(s) = \frac{b\omega}{s^2 + \omega^2}$$

$$f(t) = b^{-at} \cos \omega t \to F(s) = \frac{b(s + a)}{(s + a)^2 + \omega^2}$$

$$f(t) = b^{-at} \sin \omega t \to F(s) = \frac{b\omega}{(s + a)^2 + \omega^2}$$

Note that in all of these examples, the LT $F(s)$ is a ratio of polynomials in s. For each function $f(t)$ with LT $F(s)$ given here, it is also important to note that $f(t)$ is the inverse Laplace transform of $F(s)$. The LTs of a large collection of functions can be generated from the transforms given by using linearity of the LT. Linearity means that if $F(s)$ is the LT of $f(t)$ and $G(s)$ is the LT of $g(t)$, then for any scalars a and b, the LT of $af(t) + bg(t)$ is equal to $aF(s) + bG(s)$.

Another property that we need is that the LT of the derivative $df(t)/dt$ of a function $f(t)$ is equal to $sF(s)$ if $f(0) = 0$, where $F(s)$ is the LT of $f(t)$. Hence, differentiation in the time domain corresponds to multiplication by s in the LT domain, assuming that the initial value $f(0)$ is zero. The LT of the second derivative $d^2f(t)/dt^2$ is equal to $s^2F(s)$, assuming that $f(0) = 0$ and $\dot{f}(0) = 0$, where the "overdot" denotes the derivative.

Return To Level Control

Again consider Tank 2 and the proportional controller, with the closed-loop system given by the differential equation

$$\frac{dy(t)}{dt} + \alpha K_p y(t) = \alpha K_p r(t) \tag{2.9}$$

To solve (2.9) for the level $y(t)$, we first transform the equation into an algebraic equation by taking the LT of both sides. Assuming that $y(0) = 0$, we have

$$sY(s) + \alpha K_p Y(s) = \alpha K_p R(s) \tag{2.10}$$

where $Y(s)$ is the LT of the level $y(t)$ and $R(s)$ is the LT of the reference $r(t)$. Solving (2.10) for $Y(s)$ gives

$$Y(s) = \frac{\alpha K_p}{s + \alpha K_p} R(s) \tag{2.11}$$

Equation (2.11) is called the *closed-loop transfer function representation* and the factor

$$\frac{\alpha K_p}{s + \alpha K_p}$$

is the *closed-loop transfer function*. This particular transfer function is said to be first order since the degree of the polynomial in the denominator is equal to one. The value of s equal to $-\alpha K_p$ is called the *pole* of the transfer function, or the pole of the closed-loop system. The pole is the value of s for which the denominator of the transfer function is equal to zero. As will be seen, it plays a very significant role in the analysis of system behavior.

To solve for $y(t)$ using (2.11), we first need to determine $R(s)$. Since $r(t) = L$ for $t \geq 0$, the LT is $R(s) = L/s$. Inserting this into (2.11) gives

$$Y(s) = \frac{\alpha L K_p}{s\left(s + \alpha K_p\right)} \tag{2.12}$$

To compute $y(t)$, we need to take the inverse LT of the right-hand side of (2.12). This can be accomplished using the partial fraction expansion

$$\frac{c}{(s+a)(s+b)} = \frac{c/(b-a)}{s+a} + \frac{c/(a-b)}{s+b} \qquad (2.13)$$

where $a \neq b$. Applying (2.13) with $a = 0$, $b = \alpha K_P$, and $c = \alpha L K_P$ to the right-hand side of (2.12) gives

$$Y(s) = \frac{L}{s} - \frac{L}{s + \alpha K_P} \qquad (2.14)$$

Finally, inverse transforming (2.14) yields

$$y(t) = L - Le^{-\alpha K_P t}, \quad t \geq 0 \qquad (2.15)$$

or

$$y(t) = L(1 - e^{-\alpha K_P t}), \quad t \geq 0 \qquad (2.16)$$

For the case when $\alpha = 0.5$ in/sec, $L = 10$ in, and $K_P = 0.1$, the level $y(t)$ is plotted in Figure 2.5. The plot was generated using MATLAB 5.0 with the commands

$$g = tf(0.5,[1\ 0.05])$$

$$step(g)$$

where $0.5 = \alpha L K_P$ and $0.05 = \alpha K_P$.

From Figure 2.5, we see that the level $y(t)$ does converge to the desired level of 10 in. In fact, since $\alpha K_P > 0$, it follows directly from (2.16) that the level $y(t)$ converges to any desired level L. Hence, the proportional controller does give the desired steady-state behavior; that is, for large t, $y(t)$ is approximately equal to L.

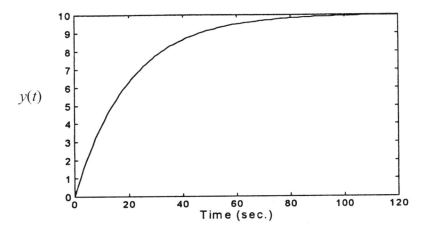

Figure 2.5 Level $y(t)$ of liquid in Tank 2 with $y(0) = 0$.

The second term $-Le^{-\alpha K_p t}$ on the right-hand side of (2.15) is called the *transient part* of the response since it decays to zero as $t \to \infty$. The rate at which the exponential transient decays to zero is characterizable in terms of the *time constant* τ, where τ is the value of t for which the transient is equal to $0.37L$. Another way to define the time constant is that it is the value of t for which $y(t)$ is equal to 63% of the steady-state value L. To compute τ, we set $y(\tau) = 0.63L$ in (2.16), which gives

$$0.63L = L(1 - e^{-\alpha K_p \tau})$$

Solving for τ yields

$$\tau = \frac{1}{\alpha K_P} \qquad (2.17)$$

For the response shown in Figure 2.5, the time constant is

$$\tau = \frac{1}{(0.5)(0.1)} = 20 \text{ seconds}$$

Hence, in this case, at time $t = 20$ sec. the level is at 6.3 in, which is 63% of the desired level of 10 in.

It is important to note that the smaller the time constant τ is, the faster the transient is; that is, the faster the transient decays to zero. Thus from (2.17), we see that a suitably fast transient is achievable by taking the gain K_P of the proportional controller to be sufficiently large. Note that this is equivalent to making the closed-loop pole $-\alpha K_P$ sufficiently negative. However, in this problem K_P cannot be chosen to be arbitrarily large due to the constraint that the value of the control input $iv(t)$ must be less than or equal to one. To see how small the time constant τ can be set in this problem, we first consider the error $e(t) = L - y(t)$. Inserting the expression (2.16) for $y(t)$, we have

$$e(t) = L - y(t) = (Le^{-\alpha K_P \tau}) \tag{2.18}$$

It follows from (2.18) that the maximum value of $e(t)$ is equal to L.

Now by definition of the proportional controller,

$$iv(t) = K_P e(t)$$

and since the maximum value of $e(t)$ is equal to L, it follows that $iv(t)$ is always less than or equal to $K_P L$. Hence, the constraint $iv(t) \leq 1$ is satisfied if $K_P L \leq 1$, or

$$K_P \leq \frac{1}{L} \tag{2.19}$$

Inserting (2.19) into the expression for the time constant τ results in the smallest possible value of τ given by

$$\tau = \frac{L}{\alpha}$$

When $\alpha = 0.5$ in/sec and $L = 10$ in, the smallest possible value of the time constant is

$$\tau = \frac{10}{0.5} = 20 \text{ seconds}$$

This value for the time constant corresponds to $K_P = 1/10 = 0.1$, which is the maximum possible value of the gain K_P of the proportional controller. This is the case considered earlier, where the level $y(t)$ is plotted in Figure 2.5.

Effect Of Output Flow

We now want to consider the effect of output flow on the proportional controller designed earlier. When the output valve position $ov(t)$ is not zero, the closed-loop system is given by the differential equation (2.7), which is repeated here:

$$\frac{dy(t)}{dt} + \alpha K_P y(t) = \alpha K_P r(t) - \beta ov(t) \qquad (2.20)$$

Taking the LT of both sides of (2.20) and solving for $Y(s)$, we have

$$Y(s) = \frac{\alpha K_P}{s + \alpha K_P} R(s) - \frac{\beta}{s + \alpha K_P} OV(s) \qquad (2.21)$$

where $OV(s)$ is the LT of $ov(t)$. We assume that the output valve is set at some fixed position so that $ov(t) = \gamma$ for $t \geq 0$, where $0 \leq \gamma \leq 1$. Then $OV(s) = \gamma/s$, and again assuming that $r(t) = L$ for $t \geq 0$, we have that $R(s) = L/s$. Inserting $OV(s)$ and $R(s)$ into (2.21) gives

$$Y(s) = \frac{\alpha L K_P}{s(s + \alpha K_P)} - \frac{\beta \gamma}{s(s + \alpha K_P)} \qquad (2.22)$$

Using (2.13) and taking the inverse LT of (2.22), we have that the level $y(t)$ is given by

$$y(t) = L - L e^{-\alpha K_P t} - \frac{\beta \gamma}{\alpha K_P} + \frac{\beta \gamma}{\alpha K_P} e^{-\alpha K_P t} \qquad (2.23)$$

It's clear from (2.23) that the steady-state value y_{ss} of the level is given by

$$y_{ss} = L - \frac{\beta \gamma}{\alpha K_P}$$

Hence, $y(t)$ no longer converges to the desired level L, unless γ is equal to zero, which means that the output valve is closed. The steady-state error e_{ss} is given by

$$e_{ss} = L - y_{ss} = \frac{\beta\gamma}{\alpha K_P}$$

To obtain some idea as to the magnitude of e_{ss}, suppose that $\alpha = \beta = 0.5$ in/sec, $K_P = 0.1$, and the output valve is half open, so that $\gamma = 0.5$. In this case the steady-state error is

$$e_{ss} = \frac{(0.5)(0.5)}{(0.5)(0.1)} = 5 \text{ in}$$

which is a sizable error. To eliminate this error, it is necessary to add integral control. We shall pursue this in the next chapter.

Problems

2.1   In the differential equation (2.5) for the tank, suppose that $\alpha = 0.5$ and the output valve is closed so that $ov(t) = 0$, in which case the equation for the tank becomes

$$\frac{dy(t)}{dt} = 0.5iv(t)$$

where $y(t)$ is the level in inches of liquid in the tank and $iv(t)$ is the position of the input valve. Assuming that $y(0) = 0$, for each of the cases given below find the Laplace transform $Y(s)$ of $y(t)$. Express $Y(s)$ as a ratio of polynomials in s.

(a) $iv(t) = 0$ for $t \geq 0$.

(b) $iv(t) = 1$ for $t \geq 0$.

(c) $iv(t) = e^{-t}$ for $t \geq 0$.

(d) $iv(t) = 1 - e^{-t}$ for $t \geq 0$.

(e) $iv(t) = (1 + \cos t)/2$ for $t \geq 0$.

2.2   Using your result in Problem 2.1:

(a)   Derive an expression for $y(t)$ for $t \geq 0$ for each of the cases given in Problem 2.1.

(b)   Using MATLAB, plot $y(t)$ for each of the cases in Problem 2.1.

(c)   Assuming that the tank is 24 inches in height, for which of the cases given in Problem 2.1 does the tank overflow?

(d)   For each case where the tank overflows, determine the time when the overflow occurs.

2.3   Now suppose that the output valve is not necessarily closed and $\beta = 0.5$, so that the differential equation for the tank becomes

$$\frac{dy(t)}{dt} = 0.5iv(t) - 0.5ov(t)$$

Assuming that $y(0) = 0$, for each of the cases given below find the Laplace transform $Y(s)$ of $y(t)$. Express $Y(s)$ as a ratio of polynomials in s.

(a) $iv(t) = 1$ for $t \geq 0$, $ov(t) = 1$ for $t \geq 0$.

(b) $iv(t) = 1$ for $t \geq 0$, $ov(t) = 1 - e^{-t}$ for $t \geq 0$.

(c) $iv(t) = e^{-t}$ for $t \geq 0$, $ov(t) = 1 - e^{-t}$ for $t \geq 0$.

(d) $iv(t) = 0.5$ for $t \geq 0$, $ov(t) = (1 + \cos t)/2$ for $t \geq 0$.

2.4   Using your result in Problem 2.3:

(a)   Derive an expression for $y(t)$ for $t \geq 0$ for each of the cases given in Problem 2.3

(b)   Using MATLAB, plot $y(t)$ for each of the cases in Problem 2.3.

(c)   Assuming that the tank is 24 inches in height, for which of the cases given in Problem 2.3 does the tank overflow?

2.5   For the tank given by the differential equation

$$\frac{dy(t)}{dt} = 0.5iv(t)$$

suppose that we use proportional control so that $iv(t) = K_P e(t) = K_P[r(t) - y(t)]$ where $e(t)$ is the tracking error and $r(t)$ is the reference input. With the proportional controller, the differential equation for the tank is

$$\frac{dy(t)}{dt} = -0.5K_P y(t) + 0.5K_P r(t)$$

(a) Using the Laplace transform, derive an expression for $y(t)$ when $y(0) = 0$, $r(t) = L = 10$ in for $t \geq 0$, and
   (i) $K_P = 0.01$
   (ii) $K_P = 0.05$
   (iii) $K_P = 0.1$

(b) For each of the cases in part (a), plot $y(t)$ on the same graph using MATLAB.

(c) Compare the responses obtained for the three different cases. Explain the differences in the responses and the reasons for the differences.

2.6 For each of the three cases in Problem 2.5, use the Laplace transform to derive an expression for $iv(t)$ and plot the results on the same graph using MATLAB.

2.7 Add the output valve term (with $\beta = 0.5$) to the equation for the tank with the proportional controller, so that the differential equation becomes

$$\frac{dy(t)}{dt} = -0.5K_P y(t) + 0.5K_P r(t) - 0.5ov(t)$$

(a) Using the Laplace transform, derive an expression for $y(t)$ when $y(0) = 0$, $K_P = 0.1$, $r(t) = 10$ in for $t \geq 0$, and
   (i) $ov(t) = 0.1$ for $t \geq 0$
   (ii) $ov(t) = 0.5$ for $t \geq 0$
   (iii) $ov(t) = e^{-t}$ for $t \geq 0$

(iv) $ov(t) = (1 + \cos t)/2$ for $t \geq 0$

(b) For each of the four cases in part (a), plot $y(t)$ on the same graph using MATLAB.

(c) Compare the responses obtained in part (b) for the different $ov(t)$. What do you conclude?

2.8 A gas furnace in a manufacturing facility has a valve that controls the flow of gas into the furnace. The position of the valve at time t is denoted by $v(t)$, where $v(t)$ ranges over the values $0 \leq v(t) \leq 1$. The valve is closed when $v(t) = 0$ and it is completely open when $v(t) = 1$. The temperature $T(t)$ inside the furnace is given by

$$\frac{dT(t)}{dt} = 2v(t) - \beta(t)$$

where $-\beta(t)$ is the temperature decrease at time t in deg/min due to heat loss.

(a) Design a proportional controller given by $v(t) = K_P[r(t) - T(t)]$, so that the temperature $T(t)$ has the fastest possible response time when $T(0) = 0$, $r(t) = 200°$ for $t \geq 0$, and $\beta(t) = 0$. Express your answer by giving the value of K_P.

(b) For your design found in part (a), derive an expression for the response $T(t)$ and plot it using MATLAB when $T(0) = 0$, $r(t) = 200°$ for $t \geq 0$, and $\beta(t) = 0$ for $t \geq 0$.

(c) For your design found in part (a), derive an expression for the response $T(t)$ and plot it using MATLAB when $T(0) = 0$, $r(t) = 200°$ for $t \geq 0$, and $\beta(t) = 0.5$ for $t \geq 0$.

2.9 A time-driven continuous-variable system with control input $c(t)$ and output $y(t)$ is given by the differential equation

$$\frac{dy(t)}{dt} = ay(t) + bc(t)$$

where a and b are constants. With the proportional control $c(t) = K_P[r(t) - y(t)]$, the closed-loop differential equation is

$$\frac{dy(t)}{dt} = [a - bK_P]y(t) + bK_P r(t)$$

(a) Derive the closed-loop transfer function representation relating $R(s)$ and $Y(s)$ where $R(s)$ and $Y(s)$ are the Laplace transforms of $r(t)$ and $y(t)$, respectively.

(b) Give the closed-loop transfer function.

(c) What is the value of the pole of the closed-loop transfer function.

(d) Using your result in part (a), derive an expression for the output response $y(t)$ when $y(0) = 0$ and $r(t) = r_0$ for $t \geq 0$, where r_0 is a constant.

(e) Using MATLAB and your result in part (d), plot on the same graph the output responses $y(t)$ when $a = 2$, $b = 1$, $r(t) = 1$ for $t \geq 0$, and
(i) $K_P = 0$
(ii) $K_P = 0.1$
(iii) $K_P = 1.0$
(iv) $K_P = 10$

(f) Compare your results obtained in part (e). Does $y(t)$ converge to 1? What do you conclude?

2.10 In Problem 2.9, now suppose that $y(0) = y_0 \neq 0$.

(a) Using the result that the Laplace transform of $dy(t)/dt$ is equal to $sY(s) - y_0$, derive an expression for the output response $y(t)$ of the closed-loop system when $y(0) = y_0$ and $r(t) = r_0$ for $t \geq 0$, where r_0 is a constant.

(b) Use MATLAB to plot your result in part (a) when $y(0) = 0.5$, $r_0 = 1$, $a = 2$, $b = 1$, and $K_P = 1$.

(c) What is the effect of having $y(0) \neq 0$. Explain.

Chapter 3

Modeling of Continuous-Variable Processes

In this chapter we consider the transfer function model for a continuous-variable (time-driven) process or system with a single input $c(t)$ and single output $y(t)$. We show how the transfer function model can be generated from a differential equation model of the process, or from experimental data describing the behavior of the process. Modeling is very important since the control approach to be presented in Chapters 4 and 5 is based on having a model of the process.

Throughout the chapter we assume that the process under consideration is *linear* over the region of operation. For example, the region of operation may consist of all process inputs $c(t)$ whose magnitude $|c(t)|$ is constrained to be less than or equal to some fixed constant M. For operation under this constraint, linearity of the process means that if $y_1(t)$ and $y_2(t)$ are the responses to inputs $c_1(t)$ and $c_2(t)$ with $|c_1(t)| \leq M$ and $|c_2(t)| \leq M$, then for any scalars α and β such that

$$|\alpha c_1(t) + \beta c_2(t)| \leq M$$

the response to the input $\alpha c_1(t) + \beta c_2(t)$ is equal to $\alpha y_1(t) + \beta y_2(t)$. Here it is assumed that the system is "at rest" before the application of an input; that is, the output and its derivatives are equal to zero before the input is applied.

An example of a process with a constrained input is the filling of a tank with the position $iv(t)$ of the input valve constrained to have values between 0 and 1. As discussed in Chapter 2, the process is given by the differential equation

$$\frac{dy(t)}{dt} = \alpha iv(t) \qquad (3.1)$$

where $y(t)$ is the level at time t of the fluid in the tank. Since (3.1) is a linear differential equation, it follows that the process given by (3.1) is linear over the region of operation where $iv(t)$ is constrained to have values between 0 and 1. The verification that the property of linearity holds is left to the reader.

In this chapter, we also assume that the process under consideration is *time invariant or constant*. This means that if $y(t)$ is the response to $c(t)$, then for any constant t_o, the response to the time shifted input $c(t - t_o)$ is equal to the time shifted output $y(t - t_o)$. Again, it is assumed that the system is at rest before an input is applied.

An example of a constant process is the tank given by (3.1). Time invariance in this case is a result of the coefficient α being a constant. If α were a function of t, the process would be *time varying*, which means that the process is not constant. The reader is invited to verify that the process given by (3.1) is constant.

Transfer Function Representation

A process or system that is both linear and constant can be described in terms of its *transfer function representation* given by

$$Y(s) = G(s)C(s) \qquad (3.2)$$

where $Y(s)$ is the Laplace transform (LT) of the output $y(t)$, $C(s)$ is the LT of the input $c(t)$, and $G(s)$ is the *transfer function* of the process. In the representation (3.2), it is assumed that the input $c(t)$ is applied to the system during the time interval $t \geq 0$ and the system is at rest before the application of $c(t)$.

From (3.2), we see that the LT $Y(s)$ of the output is equal to the product of the transfer function $G(s)$ with the LT $C(s)$ of the input. We can then divide both sides of (3.2) by $C(s)$, which yields the following expression for the transfer function:

$$G(s) = \frac{Y(s)}{C(s)} \qquad (3.3)$$

If the response $y(t)$ of a linear constant system to input $c(t)$ is known, then the system's transfer function can be determined using (3.3), assuming that the LTs of $y(t)$ and $c(t)$ can be computed. The determination of $G(s)$ using (3.3) is considered in Problem 3.3.

The transfer function representation (3.2) can be determined by applying the LT to a differential equation representation of the process. For example, suppose that the process under study is given by the first-order differential equation

$$\frac{dy(t)}{dt} + ay(t) = bc(t) \tag{3.4}$$

where a and b are constants. Taking the LT of both sides of (3.4), we have that

$$sY(s) + aY(s) = bC(s)$$

where it is assumed that $y(0) = 0$. Then solving for $Y(s)$ gives the transfer function representation

$$Y(s) = \frac{b}{s + a}C(s)$$

Hence, in this case the transfer function is

$$G(s) = \frac{b}{s + a}$$

This transfer function is said to be first order since the degree of the polynomial in the denominator is equal to one. For the tank given by the first-order differential equation (3.1), we see that $a = 0$ and $b = \alpha$, and thus the transfer function of the tank is given by

$$G(s) = \frac{\alpha}{s}$$

For another example, suppose that the process is given by the second-order differential equation

$$\frac{d^2 y(t)}{dt^2} + a_1 \frac{dy(t)}{dt} + a_0 y(t) = b_1 \frac{dc(t)}{dt} + b_0 c(t) \tag{3.5}$$

where a_1, a_0, b_1, b_0 are constants. Transforming (3.5) and solving for $Y(s)$ yields the transfer function representation $Y(s) = G(s)C(s)$ with the transfer function $G(s)$ given by

$$G(s) = \frac{b_1 s + b_0}{s^2 + a_1 s + a_0}$$

In this case, the transfer function is second order.

An example of a system with a second-order transfer function is a field-controlled dc motor with a load attached to the motor shaft. The motor with load is given by the differential equation

$$\frac{d^2 \theta(t)}{dt^2} + \frac{k_d}{I} \frac{d\theta(t)}{dt} = \frac{k_m}{I} v(t) \tag{3.6}$$

where k_d is the viscous friction coefficient of the motor and load, I is the moment of inertia of the motor and load, k_m is the motor constant, $\theta(t)$ is the angle of the motor shaft, and $v(t)$ is the voltage applied to the field circuit of the motor. Comparing (3.5) and (3.6) we see that

$$a_1 = \frac{k_d}{I}, \quad a_0 = 0, \quad b_1 = 0, \quad \text{and} \quad b_0 = \frac{k_m}{I}$$

and thus the transfer function of the motor is

$$G(s) = \frac{k_m/I}{s\left(s + \frac{k_d}{I}\right)}$$

In the examples given, the transfer function is a ratio of polynomials in s with real coefficients. In the general case when the process is *finite dimensional* or *lumped*, the transfer function $G(s)$ is a ratio of polynomials $N(s)$ and $D(s)$ in s given by

$$G(s) = \frac{N(s)}{D(s)} = \frac{b_m s^m + b_{m-1} s^{m-1} + \cdots + b_1 s + b_0}{s^n + a_{n-1} s^{n-1} + \cdots + a_1 s + a_0} \qquad (3.7)$$

where m is a nonnegative integer, n is a positive integer, and the a_i and b_i are constants. The integer n is called the *order* of the transfer function or the order of the process. The values of s for which $N(s) = 0$ are called the zeros of $G(s)$, or the zeros of the system, and the values of s for which $D(s) = 0$ are called the *poles* of $G(s)$, or the poles of the system. Hence, the zeros are the values of s for which $G(s) = 0$, and the poles are the values of s for which $G(s) = \infty$. Note that the tank system defined above has no zeros and one pole at $s = 0$, and the motor has no zeros and two poles at $s = 0$ and $s = -k_d/I$.

As shown by the examples given above, if the process under consideration is specified by a differential equation, the transfer function $G(s)$ can be computed by applying the LT. However, in practice it is often the case that the transfer function must be determined from experimental data describing the behavior of the process. As shown later, one way to accomplish this is to consider the step response of the process.

System Modeling Based On The Step Response

Given a linear constant continuous-variable process with input $c(t)$ and output $y(t)$, the *step response* of the process is the output $y(t)$ when the input $c(t)$ is a step function given by $c(t) = c_o$ for $t \geq 0$, where c_o is a constant, and the process is at rest before the step is applied. When $c_o = 1$, the response is called the *unit-step response*. Since the LT of $c(t) = c_o$ for $t \geq 0$ is equal to c_o/s, the LT $Y(s)$ of the step response $y(t)$ is given by

$$Y(s) = G(s) \frac{c_o}{s}$$

where $G(s)$ is the transfer function of the process.

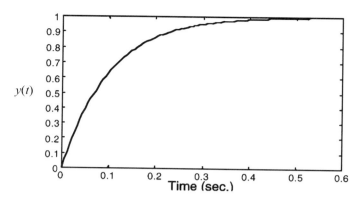

Figure 3.1   Step response of first-order process

Now suppose that the process is given by the first-order transfer function $G(s) = b/(s + a)$. Then in this case, the LT $Y(s)$ of the step response is given by

$$Y(s) = \frac{bc_0}{s(s + a)} \tag{3.8}$$

Expanding the right-hand side of (3.8) using (2.13) gives

$$Y(s) = \frac{bc_0/a}{s} - \frac{bc_0/a}{s + a} \tag{3.9}$$

and inverse transforming the right-hand side of (3.9) yields the step response:

$$y(t) = \frac{bc_0}{a} - \frac{bc_0}{a}e^{-at} = \frac{bc_0}{a}\left(1 - e^{-at}\right), \quad t \geq 0 \tag{3.10}$$

From (3.10), we see that if $a > 0$, the step response $y(t)$ converges to the steady-state value bc_0/a as $t \to \infty$, and the time constant τ of the response is equal to $1/a$. Recall from the analysis in Chapter 2 that the time constant τ is the value of t for which the response is equal to 63% of the steady-state value. For the case when $c_0 = 1$ and $b = a = 10$, the steady-state value is equal to 1, the time constant is equal to 0.1 sec, and the step response is as shown in Figure 3.1. The plot in Figure 3.1 was generated using MATLAB 5.0 with the commands

$$g = \text{tf}(10,[1\ 10])$$
$$\text{step}(g)$$

Now given some process, if the response of the process to a step input $c(t) = c_o$ for $t \geq 0$ has the form given in Figure 3.1, a transfer function for the process can be generated as follows. From the plot of the step response, determine the steady-state value y_{ss} and the time constant τ. Then from the preceding analysis, we see that the step response can be approximated by

$$y(t) \approx y_{ss}\left[1 - e^{-t/\tau}\right], \quad t \geq 0 \tag{3.11}$$

Comparing (3.11) and (3.10), we have

$$\frac{bc_o}{a} = y_{ss} \quad \text{and} \quad a = \frac{1}{\tau}$$

Solving for b gives $b = y_{ss}/c_o\tau$, and thus the process can be modeled by the first-order transfer function

$$G(s) = \frac{y_{ss}/c_o\tau}{s + (1/\tau)} \tag{3.12}$$

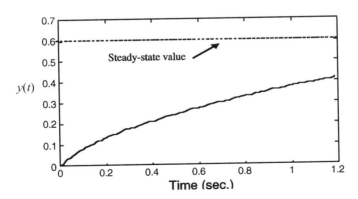

Figure 3.2   Measured unit-step response of a process.

To illustrate this construction, consider the process with the measured unit-step response shown in Figure 3.2. Since the input is a unit step, $c_o = 1$, and from the plot, we see that $y_{ss} = 0.6$. Using a ruler, we conclude that $y(t) = 0.63(0.6) = 0.378$ when $t \approx 1.0$ sec, and thus the time constant is approximately equal 1.0 sec. Hence, this process can be modeled by the transfer function

$$G(s) = \frac{0.6}{s + 1}$$

To check the accuracy of this model, its unit-step response is plotted in Figure 3.3 along with the original step response given in Figure 3.2. As seen from the plots, the two step responses are very close, and thus the model should provide a good approximation to the process.

Unbounded Step Response

If the step response $y(t)$ of a process is unbounded (i.e., grows without bound as t increases), one can model the process by first computing the derivative of the step response. This assumes that the step response data are not degraded by noise; otherwise, the derivative may be "too noisy" for analysis. The derivative $\dot{y}(t)$ may be computed using the relationship

$$\dot{y}(nT) = \frac{y(nT + T) - y(nT)}{T}$$

where T is the sampling interval and n is integer valued.

If the derivative $\dot{y}(t)$ has the form given in Figure 3.1, the transfer function $H(s)$ in the relationship $\dot{Y}(s) = H(s)C(s)$, where $\dot{Y}(s)$ is the LT of $\dot{y}(t)$, can be approximated by the procedure given above. Then since $\dot{Y}(s) = sY(s)$, we have that

$$sY(s) = H(s)C(s) \quad \text{or} \quad Y(s) = \frac{1}{s}H(s)C(s)$$

But $Y(s) = G(s)C(s)$, where $G(s)$ is the transfer function of the process and thus, $G(s)$ is equal to $1/s$ times $H(s)$. See Problem 3.5 for an application of this construction.

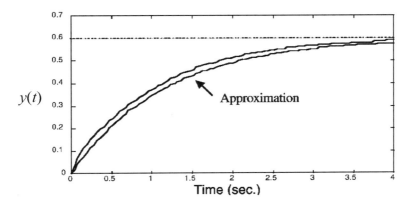

Figure 3.3 Original and model unit-step responses.

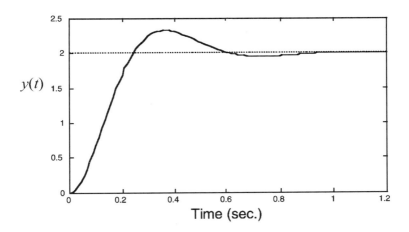

Figure 3.4 Step response with overshoot.

Higher Order Models

If the step response of a process converges to a steady-state value y_{ss}, but the response has the form shown in Figure 3.4, then in this case the process cannot be modeled by a first-order transfer function. The reason is that the step response shown in Figure 3.4 has an overshoot; whereas, there is no overshoot in a first-order response. Hence, when there is overshoot in the step response, the process must be modeled by a second-order or higher order transfer function. A process with the step response shown in Figure 3.4 can be modeled by the following second-order transfer function:

$$G(s) = \frac{y_{ss}\omega_n^2}{s^2 + 2\zeta\omega_n s + \omega_n^2} \tag{3.13}$$

where y_{ss} is the steady-state value, the constant ζ is called the *damping ratio*, and the constant ω_n is called the *natural frequency*.

For a process with transfer function (3.13), when $0 \le \zeta < 1$ the transient part of the step response contains sinusoidal terms that can cause an overshoot in the step response. This is called the *underdamped case* (i.e., when $0 \le \zeta < 1$). In the underdamped case, the two poles of $G(s)$ are complex. They can be computed by "completing the square" in the denominator of $G(s)$, which gives

$$G(s) = \frac{y_{ss}\omega_n^2}{\left(s + \zeta\omega_n\right)^2 + \omega_d^2} \tag{3.14}$$

where ω_d is the *damped natural frequency* given by $\omega_d = \omega_n\sqrt{1 - \zeta^2}$. It follows from (3.14) that when $0 \le \zeta < 1$, the poles of $G(s)$ are a complex conjugate pair with the values $s = -\zeta\omega_n + j\omega_d$ and $s = -\zeta\omega_n - j\omega_d$, where $j = \sqrt{-1}$.

It can be shown that the unit-step response $y(t)$ of the process with the transfer function (3.14) is given by

$$y(t) = y_{ss}\left[1 - e^{-\zeta\omega_n t}\left(\cos \omega_d t + \frac{\zeta\omega_n}{\omega_d} \sin \omega_d t\right)\right], \quad t \ge 0 \tag{3.15}$$

where again it is assumed that $0 \le \zeta < 1$. It follows from (3.15) that the *peak time* t_p when the step response $y(t)$ has its peak value is given by

$$t_p = \frac{\pi}{\omega_d} \tag{3.16}$$

and the *settling time* t_s when the step response reaches 5% of the steady-state value y_{ss} is given approximately by

$$t_s = \frac{3}{\zeta \omega_n} \tag{3.17}$$

Now (3.16) and (3.17) can be solved for ζ and ω_n as follows: First, from (3.17) we have that

$$\zeta = \frac{3}{t_s \omega_n} \tag{3.18}$$

Then squaring both sides of (3.16) and using the relationship

$$\omega_d = \omega_n \sqrt{1 - \zeta^2}$$

we have that

$$\frac{\pi^2}{t_p^2} = \omega_d^2 = \omega_n^2 \left(1 - \zeta^2\right) \tag{3.19}$$

Squaring both sides of (3.18) and inserting into (3.19) gives

$$\frac{\pi^2}{t_p^2} = \omega_n^2 - \frac{9}{t_s^2}$$

and thus

$$\omega_n = \sqrt{\frac{\pi^2}{t_p^2} + \frac{9}{t_s^2}} \tag{3.20}$$

Hence, using (3.20) and (3.18), we can determine the parameters ζ and ω_n from the peak time t_p and the settling time t_s.

To illustrate the computation, we fit the step response shown in Figure 3.4 to a transfer function of the form (3.13). From the plot in Figure 3.4, we have that $y_{ss} = 2$, $t_p = 0.35$ sec., and $t_s = 0.55$ sec. Inserting the values for t_p and t_s into (3.18) and (3.20) results in the values $\zeta = 0.52$ and $\omega_n = 10.6$. Hence, the transfer function is

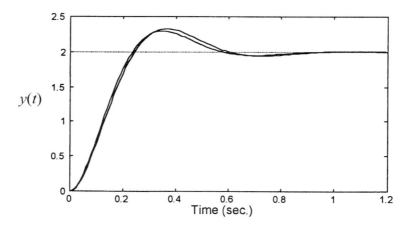

Figure 3.5   Original and model unit-step responses

$$G(s) = \frac{2(10.6^2)}{s^2 + 2(0.52)(10.6)s + 10.6^2}$$

$$= \frac{224.7}{s^2 + 11.0s + 112.4} \qquad (3.21)$$

The original unit-step response shown in Figure 3.4 and the unit-step response of the process modeled by the transfer function (3.21) are plotted in Figure 3.5. As seen from the plot, the step responses are very close, and thus the transfer function provides an accurate model of the process.

Input Delay

Many processes arising in practice have a time delay in the input. For example, the tank studied in Chapter 2 has a delay in the input, with the differential equation model of the tank given by

$$\frac{dy(t)}{dt} = \alpha i v(t - t_d) \qquad (3.22)$$

where again $y(t)$ is the level of fluid in the tank and $iv(t)$ is the position of the input valve. Here t_d is the time delay in the response of the input valve.

The process given by (3.22) is still linear (over the region of operation) and constant, and thus it has a transfer function representation. To find it, we can take the LT of both sides of (3.22), but now we need the result that the LT of the time shift $iv(t - t_d)$ is equal to $e^{-t_d s} IV(s)$, where $IV(s)$ is the LT of $iv(t)$. Then taking the LT of both sides of (3.22) gives

$$Y(s) = \frac{\alpha}{s} e^{-t_d s} IV(s) \tag{3.23}$$

From (3.23), we see that the transfer function is given by

$$G(s) = \frac{\alpha}{s} e^{-t_d s} \tag{3.24}$$

The transfer function $G(s)$ given by (3.24) is no longer a ratio of polynomials in s due to the exponential term. As a result, the system is not finite dimensional or lumped. In other words, the presence of a time delay makes the system infinite dimensional.

In the general case, a process with a time delay of t_d seconds in the input can be represented by a transfer function $G(s)$ of the form

$$G(s) = \frac{N(s)}{D(s)} e^{-t_d s} \tag{3.25}$$

where $N(s)$ and $D(s)$ are polynomials in s. Processes having a transfer function given by (3.25) are much more difficult to study using analysis techniques due to the exponential term. To proceed, one usually either neglects the time delay or makes a first-order *Pade approximation* of the exponential term given by

$$e^{-t_d s} \approx \frac{-s + \dfrac{2}{t_d}}{s + \dfrac{2}{t_d}} \tag{3.26}$$

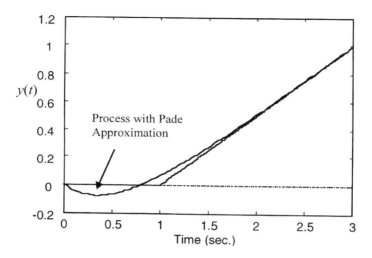

Figure 3.6   Unit-step response of exact and approximate transfer functions.

Inserting (3.26) into (3.25) results in the finite-dimensional transfer function

$$G(s) \; = \; \frac{\left(-s + \dfrac{2}{t_d}\right)N(s)}{\left(s + \dfrac{2}{t_d}\right)D(s)} \tag{3.27}$$

The process can then be studied using this transfer function. However, the use of (3.27) may not yield satisfactory results if the first-order Pade approximation is not sufficiently accurate. In such cases, one can use a higher order Pade approximation, but we do not pursue this.

   To illustrate the use of the first-order approximation (3.26), insert (3.26) into the transfer function (3.24) for the tank. The result is

$$G(s) \; = \; \frac{\alpha\left(-s + \dfrac{2}{t_d}\right)}{s\left(s + \dfrac{2}{t_d}\right)}$$

To check the accuracy of this finite-dimensional approximation, in the case when $\alpha = 0.5$ and $t_d = 1$ sec, the unit-step response for the approximation is plotted in Figure 3.6 together with the unit-step response of the system with the 1-sec time delay. The MATLAB commands are

$$g1 = \text{tf}\,(0.5,[1\ 0],\text{'td '},1)$$

$$g2 = \text{tf}\,([-0.5\ 1],[1\ 2\ 0])$$

$$\text{step}(g1,g2)$$

As seen from the plot, the approximation is off a fair amount in the neighborhood of 1 sec; however, for control purposes the approximation may be sufficiently accurate. We will investigate this in the next chapter.

Problems

3.1   Determine the transfer function for the systems given by the following differential equations where $y(t)$ is the output and $c(t)$ is the input:

(a) $\dfrac{dy(t)}{dt} + 4y(t) = c(t)$

(b) $\dfrac{dy(t)}{dt} = c(t) + \dfrac{dc(t)}{dt}$

(c) $\dfrac{d^2 y(t)}{dt^2} = c(t)$

(d) $\dfrac{d^2 y(t)}{dt^2} + y(t) = c(t) + 2\dfrac{dc(t)}{dt}$

3.2   For each of the systems in Problem 3.1, determine the poles and zeros of the transfer function.

3.3   A linear constant continuous-variable system produces the output
response $y(t) = \exp(-3t) \sin 5t$ for $t \geq 0$ when the input $c(t)$ is equal to
$\exp(-2t)$ for $t \geq 0$.

(a) Find the transfer function of the system

(b) Determine the poles and zeros of the transfer function.

(c) Using MATLAB, plot the step response of the system.

3.4   A linear constant continuous-variable process has the unit-step response
shown in Figure P3.4.

(a) Using the method given in this chapter, determine a first-order
model for the process. Express your answer by giving the transfer
function of the model.

(b) The actual transfer function of the process is given by

$$G(s) = \frac{5}{s^2 + 6s + 5}$$

Using MATLAB, compare the unit-step responses of the actual system
with the model. How close are they? Is the result expected? Explain.

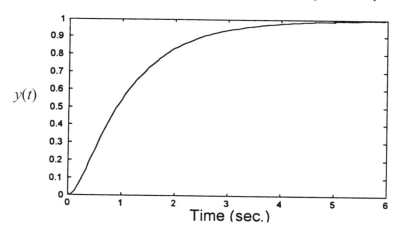

Figure P3.4

3.5   A linear constant continuous-variable process has the unit-step response
shown in Figure P3.5. Note that for a sufficiently large value of t,

$dy(t)/dt$ is equal to a constant which is the slope of the curve. Approximate $dy(t)/dt$ by

$$\frac{dy(t)}{dt} = \alpha$$

where α is the slope of the curve in Figure P3.5 for large t.

(a) Using this approximation for $dy(t)/dt$, determine a first-order model for the process. Express your answer by giving the transfer function of the model.

(b) Using MATLAB, compare the unit-step responses of the actual system with the model. How close are they? Is the result expected? Explain

3.6 A linear constant continuous-variable process has the second-order transfer function

$$G(s) = \frac{10}{s^2 + 5s + 25}$$

(a) Compute ζ, ω_n, ω_d, t_p, and t_s.

(b) Use MATLAB to plot the step response.

(c) Verify that the calculated values of t_p and t_s found in part (a) correspond to the values found from the plot in part (b).

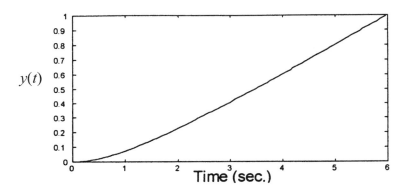

Figure P3.5

3.7   A linear constant continuous-variable process has the unit-step response
shown in Figure P3.7.

(a) Using the method given in this chapter, determine a second-order
model for the process. Express your answer by giving the transfer
function of the model.

(b) Using MATLAB, plot the step response of the model and compare
the result with the given step response. How close are they?

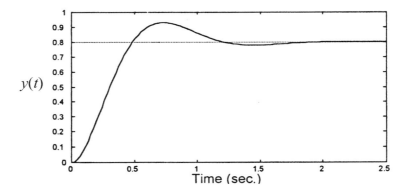

Figure P3.7

3.8   A linear constant continuous-variable process has the unit-step response
shown in Figure P3.8.

(a) Using the method given in this chapter, determine a second-order
model for the process. Express your answer by giving the transfer
function of the model.

(b) The actual transfer function of the process is given by

$$G(s) = \frac{s + 5}{s^3 + 5s^2 + 8s + 16}$$

Using MATLAB, compare the unit-step responses of the actual
system with the model. How close are they? Is the result
expected? Explain.

3.9   A linear constant continuous-variable process has the transfer function

$$G(s) \;=\; \frac{1}{s+1}\,e^{-t_d s}$$

(a) Determine the transfer function using the first-order Pade approximation of the time delay when
   (i) $t_d = 0.1$ sec
   (ii) $t_d = 0.5$ sec
   (iii) $t_d = 1$ sec
   (iv) $t_d = 2$ sec

(b) Use MATLAB to plot the step response of each of the Pade approximations in part (a).

(c) Use MATLAB to plot the exact step responses for the four cases of part (a), and compare the results with the Pade approximations found in part (a). What do you conclude? Explain.

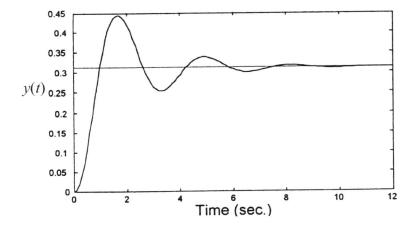

Figure P3.8

Chapter 4

Control of
Continuous-Variable Processes

Given a continuous-variable (time-driven) process with input $c(t)$ and output $y(t)$, the problem studied in this chapter is the design of a controller that forces the process output $y(t)$ to track a reference $r(t)$, where "track" means that $y(t)$ is "approximately equal" to $r(t)$. It is assumed that the process is constant and linear over the region of operation, so that it has a transfer function $G_p(s)$, where the subscript "p" on $G_p(s)$ stands for "process" or "plant." It is also assumed that the controller is linear and constant, and thus it has a transfer function $G_c(s)$, where "c" stands for "controller."

Closed-Loop Configuration

The standard closed-loop control configuration is given by the block diagram shown in Figure 4.1. As first noted in Chapter 2, there is unity feedback in the configuration since the output $y(t)$ is fed back with unity gain to the differencer where $y(t)$ is subtracted from the reference $r(t)$. In some applications the feedback loop may contain a transfer function $H(s)$ which represents the dynamics of the sensor that measures $y(t)$. However, we shall always assume that $H(s) = 1$, which is the case illustrated in Figure 4.1. With unity feedback, the signal $e(t)$ in the output of the differencer is equal to $r(t) - y(t)$, and thus $e(t)$ is the tracking error between the output and the reference. Since the control objective is to have

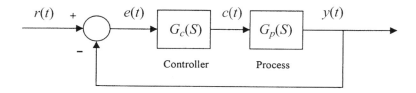

Figure 4.1 Standard feedback control configuration.

$y(t)$ track $r(t)$, this is equivalent to forcing the error $e(t)$ to be zero, or approximately zero.

The design of a controller $G_c(s)$ is carried out in terms of the closed-loop transfer function representation given by $Y(s) = G_{cl}(s)R(s)$, where $G_{cl}(s)$ is the closed-loop transfer function where the subscript "*cl*" stands for "closed loop." To determine $G_{cl}(s)$, first note that from the block diagram in Figure 4.1, we have

$$C(s) = G_c(s)E(s) \qquad (4.1)$$

$$E(s) = R(s) - Y(s) \qquad (4.2)$$

$$Y(s) = G_p(s)C(s) \qquad (4.3)$$

where $C(s)$, $Y(s)$, and $E(s)$ are the Laplace transforms of $c(t)$, $y(t)$, and $e(t)$, respectively. Inserting (4.2) into (4.1) gives

$$C(s) = G_c(s)[R(s) - Y(s)]$$

and inserting this into (4.3) yields

$$Y(s) = G_p(s)G_c(s)[R(s) - Y(s)]$$

Thus,

$$[1 + G_p(s)G_c(s)]Y(s) = G_p(s)G_c(s)R(s)$$

and solving for $Y(s)$ gives the closed-loop transfer function representation:

$$Y(s) = \frac{G_p(s)G_c(s)}{1 + G_p(s)G_c(s)} R(s) \qquad (4.4)$$

From (4.4), we see that the closed-loop transfer function $G_{cl}(s)$ is given by

$$G_{cl}(s) = \frac{G_p(s)G_c(s)}{1 + G_p(s)G_c(s)} \qquad (4.5)$$

If both the process and the controller are finite dimensional, then the transfer functions $G_p(s)$ and $G_c(s)$ are ratios of polynomials in s; that is, we have

$$G_p(s) = \frac{N_p(s)}{D_p(s)} \quad \text{and} \quad G_c(s) = \frac{N_c(s)}{D_c(s)} \qquad (4.6)$$

where $N_p(s)$, $D_p(s)$, $N_c(s)$, $D_c(s)$ are polynomials in s with real coefficients. Inserting (4.6) into (4.5), we obtain the following expression for the closed-loop transfer function:

$$G_{cl}(s) = \frac{\dfrac{N_p(s)}{D_p(s)}\dfrac{N_c(s)}{D_c(s)}}{1 + \dfrac{N_p(s)}{D_p(s)}\dfrac{N_c(s)}{D_c(s)}} = \frac{N_p(s)N_c(s)}{D_p(s)D_c(s) + N_p(s)N_c(s)} \qquad (4.7)$$

Note that if $G_p(s)$ has order n and $G_c(s)$ has order q, then the closed-loop transfer function $G_{cl}(s)$ has order $n + q$ if there is no cancellation of factors in the numerator and denominator of $G_{cl}(s)$.

The poles of the closed-loop transfer function $G_{cl}(s)$ are the values of s for which $G_{cl}(s) = \infty$, and the zeros of $G_{cl}(s)$ are the values of s for which $G_{cl}(s) = 0$. From (4.7), we see that the poles of $G_{cl}(s)$ are the values of s for which

$$D_p(s)D_c(s) + N_p(s)N_c(s) = 0$$

and the zeros of $G_{cl}(s)$ are the values of s for which

$$N_p(s)N_c(s) = 0$$

From this characterization of the poles and zeros of $G_{cl}(s)$, it follows that if $G_p(s)$ has n poles and $G_c(s)$ has q poles, then the closed-loop transfer function $G_{cl}(s)$ has $n + q$ poles if there is no cancellation of factors in the numerator and denominator of $G_{cl}(s)$.

Tracking A Step Reference

In many control problems, the reference $r(t)$ is a step function given by $r(t) = r_o$ for $t \geq 0$, where r_o is a constant called the *setpoint*. In this case, the objective is to design the controller $G_c(s)$ so that the process output $y(t)$ converges to r_o as $t \to \infty$ starting from any initial value $y(0)$. This means that we want the output response $y(t)$ to have the form

$$y(t) = r_o + y_{tr}(t), \quad t \geq 0 \tag{4.8}$$

where $y_{tr}(t) \to 0$ as $t \to \infty$. Hence, r_o is the steady-state part of the response $y(t)$ and $y_{tr}(t)$ is the *transient part* of $y(t)$. Inserting (4.8) into the tracking error $e(t) = r(t) - y(t)$, we have

$$e(t) = r_o - [r_o + y_{tr}(t)] = -y_{tr}(t), \quad t \geq 0 \tag{4.9}$$

and thus, the error $e(t)$ is equal to -1 times the transient. Note that since $y_{tr}(t) \to 0$ as $t \to \infty$, from (4.9) the steady-state error e_{ss} is equal to zero, where $e_{ss} =$ limit of $e(t)$ as $t \to \infty$.

There are three major design requirements for tracking a step reference:

1. To achieve zero steady-state error, it is necessary for $G_p(s)G_c(s)$ to have a pole at $s = 0$. This requirement follows by applying the Final Value Theorem of the LT, but we do not consider this. Later we give an alternate explanation of the requirement that there be a pole at $s = 0$.

2. The real parts of all closed-loop poles must be strictly less than zero. This is equivalent to requiring that all closed-loop poles be located in the open left-half plane (OLHP), which as shown in Figure 4.2, is that part of the complex plane located to the left of the imaginary axis. The requirement that all closed-loop poles be in the OLHP is a *stability condition*, and thus the closed-loop system is said to be stable when this property holds.

3. For the transient $y_{tr}(t)$ to be suitably fast (i.e., so that $y_{tr}(t)$ decays to zero suitably fast), the closed-loop poles must be sufficiently far over to the left of the imaginary axis in the complex plane.

We illustrate the application of these design requirements for the PI controller discussed next.

PI Controller

Suppose that the controller has the transfer function

$$G_c(s) = K_P + \frac{K_I}{s} \tag{4.10}$$

where K_P and K_I are constants. Inserting (4.10) into (4.1), we have that the LT $C(s)$ of the process control input $c(t)$ is given by

$$C(s) = G_c(s)E(s) = K_P E(s) + \frac{K_I}{s} E(s)$$

Inverse transforming results in the following expression for $c(t)$:

$$c(t) = K_P e(t) + K_I \int_{-\infty}^{t} e(\tau)\, d\tau \tag{4.11}$$

In deriving (4.11), we are using the property that if $f(t)$ has LT $F(s)$, then the inverse LT of $(1/s)F(s)$ is equal to the integral of $f(t)$. Equation (4.11) shows that the control input $c(t)$ is proportional to the tracking error $e(t)$ and the integral of the error. The first term on the right-hand-side of (4.11) is called the *proportional part* of the controller, and the second term is called the *integral part*

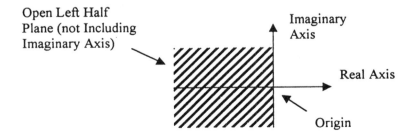

Figure 4.2 Open left-half plane.

of the controller. The controller defined by (4.11) is said to be a *proportional-plus-integral controller*, or a *PI controller*. The constant K_P is the gain of the proportional part and the constant K_I is the gain of the integral part. If $K_I = 0$, so that $G_c(s) = K_P$, we have only proportional control, and the controller is said to be a *proportional controller*, or a *P controller*.

The transfer function $G_c(s)$ of the PI controller given by (4.10) can be written in the form

$$G_c(s) = \frac{K_p s + K_I}{s} \qquad (4.12)$$

Thus, the PI controller has a zero at $s = -K_I/K_P$ and a pole at $s = 0$. The pole at $s = 0$ is a result of the integral part of the controller. Since the PI controller does have a pole at $s = 0$, by the first design requirement given earlier we see that it is possible to achieve zero steady-state error in tracking a set point by using a PI controller. Here we are assuming that the gains of the controller can be chosen so that the closed-loop system is stable, per requirement 2 given earlier.

To understand how integral control achieves zero steady-state error, consider the relationship (4.11) between the control input $c(t)$ and the tracking error $e(t)$. If the steady-state value e_{ss} of $e(t)$ is not equal to zero, then the integral term in (4.11) will grow without bound as $t \rightarrow \infty$, and as a result, $c(t)$ will grow without bound as $t \rightarrow \infty$. But if the closed-loop system is stable, $c(t)$ must be bounded; that is, the magnitude $|c(t)|$ must be less than some positive constant M for all t, and thus it must be true that $e_{ss} = 0$.

Given a process with the transfer function

$$G_p(s) = \frac{N_p(s)}{D_p(s)}$$

and the PI controller with transfer function (4.12), using (4.7) we have that the closed-loop transfer function is

$$G_{cl}(s) \;=\; \frac{N_p(s)(K_p s + K_I)}{D_p(s)s \;+\; N_p(s)(K_p s + K_I)} \qquad (4.13)$$

From (4.13) we see that the closed-loop transfer function has a zero at $s = -K_I/K_P$, which comes from the zero of the PI controller. As shown later, we can

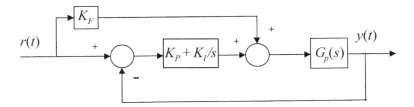

Figure 4.3 Feedback configuration with modified PI controller.

modify the PI controller so that this zero can be assigned any desired value independent of K_P and K_I.

Modified PI Controller

Consider the modified PI controller shown in Figure 4.3. Note that if the gain K_F is equal to zero, the configuration in Figure 4.3 reduces to the standard closed-loop format with a PI controller. We now determine the closed-loop transfer function for the feedback configuration in Figure 4.3.

From Figure 4.3, we have

$$Y(s) = G_p(s)\{K_F R(s) + (K_P + K_I/s)[R(s) - Y(s)]\}$$

Combining the terms involving $Y(s)$ gives

$$[1 + G_p(s)(K_P + K_I/s)]Y(s) = G_p(s)(K_F + K_P + K_I/s)R(s)$$

and solving for $Y(s)$ yields

$$Y(s) = \frac{G_p(s)(K_F + K_P + K_I/s)}{1 + G_p(s)(K_P + K_I/s)} R(s) \qquad (4.14)$$

Finally, setting $G_p(s) = N_p(s)/D_p(s)$ in (4.14) and multiplying the right-hand side by $D_p(s)s/D_p(s)s$ results in

$$Y(s) = \frac{N_p(s)[(K_F + K_P)s + K_I]}{D_p(s)s + N_p(s)(K_P s + K_I)} R(s)$$

Hence, the closed-loop transfer function $G_{cl}(s)$ is given by

$$G_{cl}(s) = \frac{N_p(s)[(K_F + K_P)s + K_I]}{D_p(s)s + N_p(s)(K_P s + K_I)} \qquad (4.15)$$

Note that the closed-loop transfer function (4.15) reduces to (4.13) when $K_F = 0$. This confirms that the modified PI controller gives the same result as the standard PI controller when $K_F = 0$. From (4.15), see that the zero due to the controller is given by

$$s = -\frac{K_I}{K_F + K_P}$$

Clearly, by choosing the gain K_F this zero can be assigned any desired value independent of the values K_P and K_I. Later it is shown that this capability is very useful in obtaining a desired control performance.

From (4.15), we see that the closed-loop poles are the zeros of the polynomial $D_p(s)s + N_p(s)(K_P s + K_I)$. A key part of the controller design is determining where the closed-poles can be placed by choosing the gains K_P and K_I. In particular, we would like to know how far over to the left they can be placed in the OLHP. We pursue this by beginning with a first-order process.

PI Control of a First-Order Process

Suppose the process has the first-order transfer function $G_p(s) = b/(s+a)$, where a and b are constants. Then with the standard PI controller $G_c(s) = K_P + K_I/s$, using (4.13) we have that the closed-loop transfer function is

$$G_{cl}(s) = \frac{bK_P s + bK_I}{(s+a)s + bK_P s + bK_I}$$

Rewriting the denominator of $G_{cl}(s)$ yields

$$G_{cl}(s) = \frac{bK_P s + bK_I}{s^2 + (a + bK_P)s + bK_I} \qquad (4.16)$$

From (4.16) we see that the closed-loop transfer function has a zero at $s = -K_I/K_P$, and it has two poles. It also follows from (4.16) that the two poles of $G_{cl}(s)$ can be placed wherever desired in the OLHP by choosing the controller gains K_P and K_I. In particular, suppose that we want the closed-loop poles to be $s = -p_1$ and $s = -p_2$. If p_1 is complex, then p_2 is complex and is equal to the complex conjugate of p_1. Now for $s = -p_1$ and $s = -p_2$ to be the closed-loop poles, it must be true that

$$s^2 + (a+bK_P)s + bK_I = (s + p_1)(s + p_2)$$

$$= s^2 + (p_1 + p_2)s + p_1 p_2$$

Hence, equating coefficients gives

$$a + bK_P = p_1 + p_2 \quad \text{and} \quad bK_I = p_1 p_2$$

which implies that

$$K_P = \frac{(p_1 + p_2) - a}{b} \quad \text{and} \quad K_I = \frac{p_1 p_2}{b} \qquad (4.17)$$

Then for any desired closed-loop poles $-p_1$ and $-p_2$, the controller gains K_P and K_I are determined using (4.17).

If the closed-loop poles $-p_1$ and $-p_2$ are chosen to be negative real numbers with $p_1 \neq p_2$, then the unit-step response of the closed-loop system will have the form

$$y(t) = 1 + c_1 e^{-p_1 t} + c_2 e^{-p_2 t}, \quad t \geq 0 \qquad (4.18)$$

where c_1 and c_2 are real constants. The transient part of the unit-step response consists of the second and third terms on the right-hand side of (4.18). If the closed-loop poles p_1 and p_2 are a complex pair given by $p_1 = -\sigma + j\omega$ and $p_2 = -\sigma - j\omega$, where $\sigma > 0$, then the unit-step response of the closed-loop system will have the form

$$y(t) = 1 + ce^{-\sigma t} \cos(\omega t + \theta), \quad t \geq 0$$

where c and θ are real constants. The transient part of this unit-step response is the cosine term with the decaying exponential.

Due to the zero of $G_{cl}(s)$ at $s = -K_I/K_P$, the transient may have overshoot even in the case when p_1 and p_2 are real numbers. We illustrate this by considering control of the dc motor that was defined in Chapter 3.

PI Control of a dc Motor

The differential equation (3.6) of a dc motor with load is repeated below:

$$\frac{d^2\theta(t)}{dt^2} + \frac{k_d}{I}\frac{d\theta(t)}{dt} = \frac{k_m}{I}v(t) \qquad (4.19)$$

where k_d is the viscous friction coefficient of the motor and load, I is the moment of inertia of the motor and load, k_m is the motor constant, $\theta(t)$ is the angle of the motor shaft, and $v(t)$ is the voltage applied to the field circuit of the motor.

The objective is to control the velocity $\dot\theta(t)$ of the motor shaft; in particular, we want $\dot\theta(t) \to \dot\theta_o$ as $t \to \infty$, where $\dot\theta_o$ is the desired velocity. Taking the LT of both sides of (4.19) gives

$$s\dot\Theta(s) + \frac{k_d}{I}\dot\Theta(s) = \frac{k_m}{I}V(s) \qquad (4.20)$$

where $\dot\Theta(s)$ is the LT of $\dot\theta(t)$ and $V(s)$ is the LT of $v(t)$. Then solving (4.20), we have

$$\dot\Theta(s) = \frac{\dfrac{k_m}{I}}{s + \dfrac{k_d}{I}}V(s)$$

and thus the transfer function for velocity control is

$$G_p(s) = \frac{\dfrac{k_m}{I}}{s + \dfrac{k_d}{I}}$$

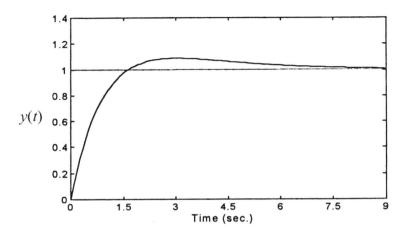

Figure 4.4 Unit-step response of motor with PI controller.

We set $I = 1$, $k_d = 0.1$, and $k_m = 5$, so that $G_p(s)$ becomes

$$G_p(s) \; = \; \frac{5}{s + 0.1}$$

Now with the PI controller $G_c(s) = K_P + K_I/s$, using (4.13) we have that the closed-loop transfer function is

$$G_{cl}(s) \; = \; \frac{5K_p s + 5K_I}{s^2 + (0.1 + 5K_P)s + 5K_I}$$

To put the closed-loop poles at $s = -0.5$ and $s = -1$, using (4.17) with $p_1 = 0.5$ and $p_2 = 1$, we have that the PI controller gains are

$$K_P = \frac{0.5 + 1 - 0.1}{5} = 0.28 \quad \text{and} \quad K_I = \frac{(0.5)(1)}{5} = 0.1$$

The unit-step response of the resulting closed-loop system is plotted in Figure 4.4. Note that even though the closed-loop poles are real, there is overshoot in the unit-step response. This turns out to be due to the zero at $s = -K_I/K_P = -0.357$ in the closed-loop transfer function $G_{cl}(s)$.

From the plot in Figure 4.4 we see that the time constant (the time to reach 63% of the steady-state value) for the response is approximately equal to 0.7 second. To achieve a faster response, we would need to select poles that are farther over in the OLHP than the values -0.5 and -1. If this is attempted, the overshoot will increase. To eliminate the overshoot, we can use the modified PI controller with the closed-loop transfer function given by (4.15). Inserting $G_p(s) = 5/(s+0.1)$ into (4.15) gives

$$G_{cl}(s) = \frac{5(K_F + K_p)s + 5K_I}{s^2 + (0.1 + 5K_p)s + 5K_I} \qquad (4.21)$$

Then given a desired time constant τ for the unit-step response, we choose K_P and K_I so that the closed-loop poles are $-p_1 = -1/\tau$ and $-p_2 = -1/\tau$, and we choose K_F so that the zero cancels one of the poles. For example, let's select $\tau = 0.5$ sec, so that $p_1 = p_2 = 2$, and

$$K_P = \frac{2 + 2 - 0.1}{5} = 0.78 \quad \text{and} \quad K_I = \frac{(2)(2)}{5} = 0.8$$

To cancel one of the poles, the zero of $G_{cl}(s)$ must equal -2, and thus

$$-\frac{K_I}{K_F + K_P} = -2$$

Solving for K_F, we have

$$K_F = \frac{K_I - 2K_P}{2} = -0.38$$

With these values for the gains, the closed-loop transfer function (4.21) becomes

$$G_{cl}(s) = \frac{2s + 4}{s^2 + 4s + 4} = \frac{2(s + 2)}{s^2 + 4s + 4} = \frac{2}{s + 2}$$

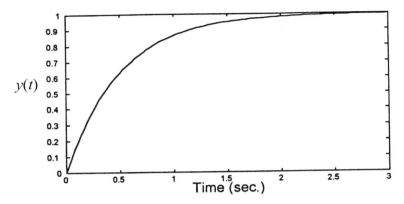

Figure 4.5   Unit-step response of closed-loop system with modified PI controller.

The unit-step response of the closed-loop system is plotted in Figure 4.5. Note that there is no overshoot (since the closed-loop system reduces to first order), and the time constant is equal to the desired value of 0.5 sec.

PID Controller

Now suppose that the goal is to control the position $\theta(t)$ of the shaft of the dc motor given by (4.19). In this case, the relevant transfer function representation is

$$\Theta(s) = \frac{\dfrac{k_m}{I}}{s\left(s + \dfrac{k_d}{I}\right)} V(s)$$

where $\Theta(s)$ is the LT of the motor shaft angle $\theta(t)$. Then with the PI controller $G_c(s) = K_P + K_I/s$, using (4.13) we have that the closed-loop transfer function is

$$G_{cl}(s) \;=\; \frac{\dfrac{k_m}{I}(K_P s + K_I)}{s^2\left(s + \dfrac{k_d}{I}\right) \;+\; \dfrac{k_m}{I}(K_P s + K_I)}$$

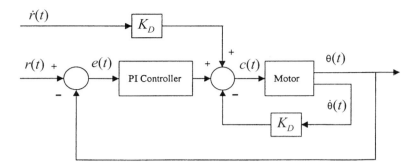

Figure 4.6 Combined PI controller and derivative controller.

$$G_{cl}(s) = \frac{\dfrac{k_m K_P}{I} s + \dfrac{k_m K_I}{I}}{s^3 + \dfrac{k_d}{I} s^2 + \dfrac{k_m K_P}{I} s + \dfrac{k_m K_I}{I}} \qquad (4.22)$$

From (4.22), see that the coefficient of s^2 in the denominator of $G_{cl}(s)$ is independent of the PI controller gains K_P and K_I. Hence, in this case it is not possible to place the closed-loop poles wherever we wish. In addition, since the viscous friction coefficient k_d is small, which implies that the coefficient k_d/I of s^2 in the denominator of $G_{cl}(s)$ is small, it turns out that it is not possible to obtain a suitably fast step response by selecting the gains K_P and K_I.

   To achieve position control of the motor shaft with a suitably fast step response, we need to add derivative control to the PI controller. This is accomplished by using the closed-loop configuration shown in Figure 4.6. In this configuration, we assume that there is a sensor that provides a measurement of the angular velocity $\dot{\theta}(t)$ of the motor shaft. The derivative part of the control action in the closed-loop system shown in Figure 4.6 is the feeding back of the angular velocity $\dot{\theta}(t)$ through the gain K_D.

   From the block diagram in Figure 4.6, we have that the LT $C(s)$ of the control signal $c(t)$ is given by

$$C(s) = \left(K_P + \frac{K_I}{s} \right) E(s) + K_D s \left[R(s) - \Theta(s) \right] = \left(K_P + \frac{K_I}{s} + K_D s \right) E(s)$$

and thus the transfer function of the combined PI controller and derivative control is

$$G_c(s) = K_P + \frac{K_I}{s} + K_D s \tag{4.23}$$

The controller with the transfer function (4.23) is called a *proportional-plus-integral-plus-derivative controller*, or *PID controller*. The derivative part of the PID controller is the term $K_D s$, where K_D is the gain of the derivative part. It is important to note that the PID controller in Figure 4.6 is implemented by using a sensor that measures the derivative of the output $\theta(t)$. If a sensor for measuring the output is not available, it is necessary to generate the derivative of the output in order to implement a PID controller. This can be a problem in some applications where the output contains noise, since differentiation will greatly amplify the noise

For the closed-loop configuration shown in Figure 4.6, we have that

$$C(s) = G_c(s)E(s)$$

$$E(s) = R(s) - \Theta(s)$$

$$\Theta(s) = G_p(s)C(s)$$

where $G_c(s)$ is the controller transfer function given by (4.23) and $G_p(s)$ is the transfer function of the motor. Solving these equations for $\Theta(s)$ gives

$$\Theta(s) = \frac{G_p(s)G_c(s)}{1 + G_p(s)G_c(s)} R(s)$$

and thus the closed-loop transfer function $G_{cl}(s)$ has the same form as before; that is,

$$G_{cl}(s) = \frac{G_p(s)G_c(s)}{1 + G_p(s)G_c(s)}$$

Inserting (4.23) and the motor transfer function

$$G_p(s) = \frac{\dfrac{k_m}{I}}{s\left(s + \dfrac{k_d}{I}\right)}$$

into $G_{cl}(s)$ yields

$$
\begin{aligned}
G_{cl}(s) &= \frac{\dfrac{(k_m/I)(K_P s + K_I + K_D s^2)}{s^2(s + k_d/I)}}{1 + \dfrac{(k_m/I)(K_P s + K_I + K_D s^2)}{s^2(s + k_d/I)}} \\[2em]
&= \frac{(k_m/I)K_D s^2 + (k_m/I)K_P s + (k_m/I)K_I}{s^2(s + (k_d/I) + (k_m/I)K_D s^2 + (k_m/I)K_P s + (k_m/I)K_I} \\[2em]
&= \frac{(k_m/I)K_D s^2 + (k_m/I)K_P s + (k_m/I)K_I}{s^3 + [(k_d/I) + (k_m/I)K_D]s^2 + (k_m/I)K_P s + (k_m/I)K_I} \quad (4.24)
\end{aligned}
$$

From (4.24) we see that the coefficients of s^2, s, and s^0 in the denominator of $G_{cl}(s)$ can be chosen to have any desired values by selecting the controller gains K_P, K_I, and K_D, and thus all three poles of the closed-loop transfer function can be placed anywhere we wish. Hence with PID control of the motor, we have complete pole assignability, and therefore it will be possible to achieve a suitably fast step response (see Problem 4.5).

Effect Of A Disturbance

In practice, a process with input $c(t)$ and output $y(t)$ is often subjected to a disturbance $d(t)$, which enters into the transfer function representation as follows:

$$Y(s) = G_p(s)C(s) + B(s)D(s) \quad (4.25)$$

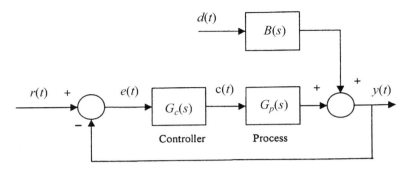

Figure 4.7 Standard feedback control configuration with disturbance.

Here $Y(s)$, $C(s)$, and $D(s)$ are the LTs of $y(t)$, $c(t)$, and $d(t)$, respectively; $G_p(s)$ is the transfer function of the process; and $B(s)$ is a constant or a ratio of polynomials in s.

For the process given by (4.25), the standard feedback control configuration is shown in Figure 4.7. Note that the LT $C(s)$ of the control input $c(t)$ is given by the same expression as in the case when there is no disturbance; that is,

$$C(s) = G_c(s)E(s) = G_c(s)[R(s) - Y(s)] \qquad (4.26)$$

The objective is to design the controller transfer function $G_c(s)$ so that the process output $y(t)$ still tracks the reference $r(t)$, even though the process is subjected to the disturbance $d(t)$. To design $G_c(s)$, we first need to determine the closed-loop transfer function representation for the configuration in Figure 4.7: Inserting (4.26) into (4.25) and solving for $Y(s)$ gives

$$Y(s) = \frac{G_p(s)G_c(s)}{1 + G_p(s)G_c(s)} R(s) + \frac{B(s)}{1 + G_p(s)G_c(s)} D(s) \qquad (4.27)$$

The first term on the right-hand side of (4.27) is the LT of the part of the output due to the reference $r(t)$ and the second term on the right-hand side of (4.27) is the LT of the part of the output due to the disturbance $d(t)$. If $d(t) = 0$ so that $D(s) = 0$, (4.27) reduces to the closed-loop transfer function representation with no disturbance; that is,

$$Y(s) = \frac{G_p(s)G_c(s)}{1 + G_p(s)G_c(s)} R(s)$$

where

$$G_{cl}(s) = \frac{G_p(s)G_c(s)}{1 + G_p(s)G_c(s)}$$

is the closed-loop transfer function. The closed-loop system defined by (4.27) is stable if all the poles of $G_{cl}(s)$ are located in the OLHP.

It is clear from (4.27) that the presence of the disturbance $d(t)$ will in general affect both the transient and steady-state responses of the closed-loop system, and as a result, the disturbance may have a major effect on the capability of having the process output $y(t)$ track the reference $r(t)$. If both $d(t)$ and $r(t)$ are step functions; that is, $r(t) = r_o$ for $t \geq 0$ and $d(t) = d_o$ for $t \geq 0$, and the controller $G_c(s)$ has a pole at zero, it follows that the steady-state value e_{ss} of the tracking error $e(t)$ is equal to zero if the closed-loop system is stable. Thus in this case the effect of the disturbance is completely rejected in steady-state operation.

As an example, again consider velocity control of a motor shaft, with the transfer function representation of the motor given by

$$\dot{\Theta}(s) = \frac{5}{s + 0.1} V(s) \tag{4.28}$$

where $\dot{\Theta}(s)$ is the LT of the angular velocity $\dot{\theta}(t)$ of the motor shaft and $V(s)$ is the LT of the voltage applied to the field circuit of the motor. If an external torque $\tau(t)$ is applied to the motor shaft while the motor is running (for example, if someone were to grab hold of the motor shaft), $\tau(t)$ will act as a disturbance applied to the motor. In this case, the transfer function representation (4.28) changes to

$$\dot{\Theta}(s) = \frac{5}{s + 0.1} V(s) + \frac{1}{s + 0.1} T(s)$$

where $T(s)$ is the LT of $\tau(t)$. Hence,

$$G_p(s) = \frac{5}{s+0.1} \quad \text{and} \quad B(s) = \frac{1}{s+0.1}$$

and if we select a PI controller $G_c(s) = K_P + K_I/s$, the transfer function representation (4.27) becomes

$$\dot{\Theta}(s) = \frac{5K_P s + 5K_I}{s^2 + (0.1 + 5K_P)s + 5K_I} R(s) + \frac{s}{s^2 + (0.1 + 5K_P)s + 5K_I} T(s)$$

If we choose the controller gains to be $K_P = 0.28$ and $K_I = 0.1$, the closed-loop poles are $s = -0.5$ and $s = -1$, and the response $y(t)$ of the closed-loop system with $r(t) = 1$ and $\tau(t) = -1$ for $t \geq 0$ is given in Figure 4.8. Also plotted in Figure 4.8 is the response when $\tau(t) = 0$. The MATLAB commands for generating these plots are

$$g1=tf([1.5\ 0.5],[1\ 1.5\ 0.5])$$

$$g2=tf([-1\ 0],[1\ 1.5\ 0.5])$$

$$step(g1,g2)$$

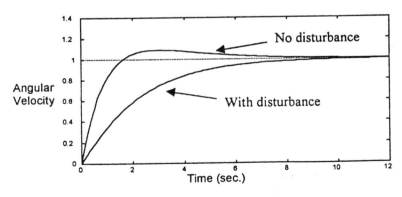

Figure 4.8 Unit-step response of closed-loop system with and without the step disturbance

In Figure 4.8 note that the negative torque resulting from the disturbance $\tau(t) = -1$ prevents the overshoot from occurring. We also see that the steady-state error in tracking the unit-step input is zero, so the step disturbance is rejected in the limit as $t \to \infty$. This is a consequence of the controller having an integral part that provides a pole at $s = 0$.

Processes With Input Delay

As discussed in Chapter 3, a process with a delay in the input can be modeled by a transfer function of the form

$$G_p(s) = \frac{N(s)}{D(s)} e^{-t_d s}$$

where $N(s)$ and $D(s)$ are polynomials in s and t_d is the time delay. Again, we want to use closed-loop control to force the process output $y(t)$ to track a reference $r(t)$. If the desired time constant of the closed-loop system is a factor of 10 or greater than the time delay t_d, then the delay can be neglected in the design of the controller transfer function $G_c(s)$. If this is not the case, the effect of the delay needs to be taken into account. We illustrate this by considering the tank level control problem first studied in Chapter 2.

Recall [see (3.23)] that when the output valve is closed, the transfer function representation of the tank is given by

$$Y(s) = \frac{\alpha}{s} e^{-t_d s} IV(s) \tag{4.29}$$

where $Y(s)$ is the LT of the level $y(t)$ of the liquid in the tank, $IV(s)$ is the LT of the position $iv(t)$ of the input valve, and t_d is the time delay in the response of the input valve. We consider proportional control, so that

$$IV(s) = K_P E(s) = K_P[R(s) - Y(s)] \tag{4.30}$$

where $R(s)$ is the LT of the reference $r(t)$ (the desired level). Inserting (4.30) into (4.29) and solving for $Y(s)$ yields the closed-loop transfer function representation

$$Y(s) = \frac{\alpha K_p e^{-t_d s}}{s + \alpha K_p e^{-t_d s}} R(s) \qquad (4.31)$$

If the delay t_d is neglected, that is, we set $t_d = 0$, then (4.31) reduces to

$$Y(s) = \frac{\alpha K_p}{s + \alpha K_p} R(s) \qquad (4.32)$$

Using (4.32) and the constraint that $0 \le iv(t) \le 1$, in Chapter 2 we found that if $\alpha = 0.5$ we could achieve a time constant of 20 sec by setting the gain K_P equal to 0.1. If the actual time delay t_d is on the order of 1 sec, the controller $G_c(s) = K_P = 0.1$ should result in good performance when the delay term is included. In other words, inserting $\alpha = 0.5$ and $K_P = 0.1$ into (4.31) should result in a unit-step response that is close to the unit-step response of the closed-loop system given by (4.32). Unfortunately, we cannot check this because the step response for the closed-loop system given by (4.31) is difficult to compute due to the presence of the exponential term in s in the denominator of the closed-loop transfer function. We can, however, check the closed-loop system step response based on an approximation of the open-loop transfer function by using a first-order Pade approximation of $e^{-t_d s}$ [see (3.26)]: Given the open-loop transfer function (with $\alpha = 0.5$ and $t_d = 1$)

$$G_p(s) = \frac{0.5}{s} e^{-s}$$

the approximation to $G_p(s)$ using the Pade approximation can be computed via the MATLAB 5.0 commands

```
g=tf(0.5,[1 0],'td',1)
pade(g,1)
```

Executing these commands results in

$$G_p(s) = \frac{-0.5s + 1}{s^2 + 2s}$$

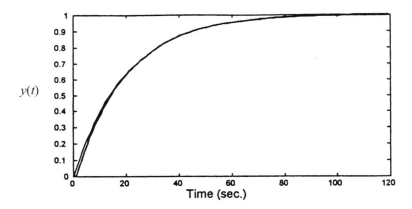

Figure 4.9   Unit-step response of closed-loop system with and without time delay.

Then with $K_P = 0.1$, the closed-loop transfer function based on this approximation is

$$G_{cl}(s) = \frac{G_p(s)C_c(s)}{1 + G_p(s)G_c(s)} = \frac{-0.05s + 0.1}{s^2 + 1.95s + 0.1}$$

The plot of the unit-step response of the closed-loop system is given in Figure 4.9. Also shown in Figure 4.9 is the step response of the closed-loop system with the delay neglected. As seen from the figure, the unit-step responses are almost identical, which verifies that the time delay can be neglected.

Now suppose that $t_d = 5$ sec. In this case the approximation to the open-loop transfer function is

$$G_p(s) = \frac{-0.5s + 0.2}{s^2 + 0.4s}$$

and with $G_c(s) = K_P = 0.1$, the closed-loop transfer function is

$$G_{cl}(s) = \frac{-0.05s + 0.02}{s^2 + 0.35s + 0.02}$$

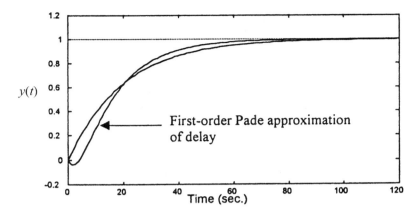

Figure 4.10 Unit-step response of closed-loop system with and without delay.

The unit-step response is plotted in Figure 4.10, along with the unit-step response of the closed-loop system with the delay neglected. As seen from the figure, there is some difference in the plots, and thus in this case neglecting the delay may result in a poor design.

Problems

4.1 Consider the process in Problem 3.4 with the unit-step response shown in Figure P3.4.

(a) Using the first-order model found in Problem 3.4, design a continuous-variable controller so that the steady-state error to a unit-step input is equal to zero and the closed-loop poles are at $s = -1, -1$. Express your answer by giving the transfer function of the controller.

(b) Repeat part (a), but now put the closed-loop poles at $s = -5, -5$.

(c) As noted in Problem 3.4, the actual transfer function of the process is

$$G_p(s) = \frac{5}{s^2 + 6s + 5}$$

Using your controller found in part (a) with the actual process transfer function, plot the unit-step response of the closed-loop system using MATLAB. Are the results as expected? Explain.

(d) Repeat part (c) using the controller found in part (b).

4.2   A linear constant continuous-variable process with input $c(t)$ and output $y(t)$ has the transfer function

$$G_p(s) = \frac{5}{s+10}$$

(a) Design a controller so that the unit-step response of the closed-loop system is equal to

$$y(t) = 1 + c_1 e^{-2t} + c_2 e^{-5t}, \quad t \geq 0$$

where c_1 and c_2 are constants. Express your answer by giving the transfer function of the controller.

(b) For the controller found in part (a), use MATLAB to plot the unit-step response of the closed-loop system.

(c) Repeats parts (a) and (b) where now the desired unit-step response is

$$y(t) = 1 + ce^{-2t}\cos(5t + \theta), \quad t \geq 0$$

where c and θ are constants. Express your answer by giving the transfer function of the controller.

4.3   Again consider a tank given by the differential equation

$$\frac{dy(t)}{dt} = 0.5iv(t) - 0.5ov(t)$$

where $y(t)$ is the level in inches of liquid in the tank, $iv(t)$ is the position of the input valve, and $ov(t)$ is the position of the output valve. Recall that $0 \leq iv(t) \leq 1$ and $0 \leq ov(t) \leq 1$.

(a) Design a modified PI controller so that whenever $ov(t)$ is equal to a constant, the steady-state error resulting from a constant reference input $r(t) = L$ is equal to zero and the response is as fast as possible. Express your answer by giving the values of the gains K_F, K_P, and K_I.

(b) For your design in part (a), use MATLAB to plot the response $y(t)$ when $y(0) = 0$, $L = 10$ in., and $ov(t) = 0.5$ for $t \geq 0$.

(c) For the values given in part (b), use MATLAB to plot the control signal $iv(t)$ that is applied to the tank. Is it true that $0 \leq iv(t) \leq 1$? If not, you need to redo your design.

4.4 A gas furnace (first considered in Problem 2.8) is given by the differential equation

$$\frac{dT(t)}{dt} = 2v(t) - \beta(t)$$

where $T(t)$ is the temperature in the furnace, $v(t)$ is the position of the valve controlling the gas into the furnace, and $-\beta(t)$ is the term due to heat loss. Recall that $0 \leq v(t) \leq 1$.

(a) Design a modified PI controller so that whenever $\beta(t)$ is equal to a constant, the steady-state error resulting from a constant reference input $r(t) = r_0$ is equal to zero and the response is as fast as possible. Express your answer by giving the values of the gains K_F, K_P, and K_I.

(b) For your design in part (a), use MATLAB to plot the response $T(t)$ when $T(0) = 0$, $r_0 = 200°$, and $\beta(t) = 0.5$ for $t \geq 0$.

(c) For the values given in part (b), use MATLAB to plot the control signal $v(t)$ that is applied to the tank. Verify that $0 \leq v(t) \leq 1$.

4.5 Consider a dc motor with transfer function

$$G_p(s) = \frac{5}{s(s + 0.1)}$$

(a) Design a proportional-plus-derivative (PD) controller given by $G_c(s) = K_P + K_D s$ so that the poles of the closed-loop system are $s = -0.5, -0.5$. Express your answer by giving the values of the gains K_P and K_D.

(b) Design a PID controller so that the poles of the closed-loop system are $s = -0.5, -0.5, -0.5$. Express your answer by giving the values of the gains K_P, K_I and K_D.

(c) For the designs in parts (a) and (b), use MATLAB to plot the unit-step responses of the closed-loop systems on the same graph. How do the responses compare? Are the results as expected? Explain.

(d) Which is the better controller (PD or PID) for the motor? Justify your answer.

4.6   Consider the process with transfer function

$$G_p(s) = \frac{1}{s(s-5)}$$

(a) Is it possible to design a PI controller for the process that results in a stable closed-loop system? If not, state why not. If so, design a PI controller that results in a stable closed-loop system.

(b) Design a PID controller so that the response of the closed-loop system is given by

$$y(t) = r_o + c_1 e^{-2t} + c_2 e^{-5t} \cos(10t + \theta), \quad t \geq 0$$

where r_o is the value of the reference input and c_1, c_2, and θ are constants.

4.7   For the motor in Problem 4.5, consider a modified PID controller given by

$$V(s) = [K_P + K_I/s + K_D s]E(s) + K_F R(s)$$

where $V(s)$, $E(s)$, and $R(s)$ are the LTs of the voltage applied to the motor, the tracking error, and the reference input, respectively.

(a) Design the modified PID controller so that the zero of the closed-loop system due to the controller is equal to -0.5 and it cancels one of the closed-loop poles and the other closed-loop poles are equal to $-0.5, -0.5$. Express your answer by giving the values of the gains K_P, K_I, K_D, and K_F.

(b) Use MATLAB to plot the unit-step response of the closed-loop system with the controller design in part (a). Compare with the unit-step responses obtained for the two designs in Problem 4.5. What do you conclude?

(c) What is gained by using the modified PID controller over the PID controller? Explain.

4.8 For the process with transfer function

$$G_p(s) = \frac{4}{s(s+10)}$$

design a modified PI controller so that the unit-step response of the closed-loop system is given by

$$y(t) = 1 + c_1 e^{-3t} + c_2 e^{-6t}, \quad t \geq 0$$

where c_1 and c_2 are constants.

4.9 Consider the process with the transfer function

$$G_p(s) = \frac{5}{s+1} e^{-s}$$

(a) Neglect the time delay and design a PI controller so that the unit-step response of the closed-loop system is given by

$$y(t) = 1 + c_1 e^{-t} + c_2 e^{-2t}, \quad t \geq 0$$

where c_1 and c_2 are constants.

(b) Using the Pade approximation of the time delay, construct an approximation of the process with the delay included. Express your answer by giving the transfer function of the approximation.

(c) Apply the PI controller designed in part (a) to the approximation of the process found in part (b). Use MATLAB to plot the unit-step response of the closed-loop system and on the same graph plot the unit-step response of the closed-loop system obtained in part (a) using the PI controller with the delay neglected. Compare the results. What do you conclude?

(d) Repeat parts (b) and (c) with the process delay equal to 5 sec.

Chapter 5

Digital Control

The P, PI, and PID controllers studied in the last chapter are examples of continuous-time, or analog, controllers that can be directly implemented using a
circuit containing operational amplifiers. However, today continuous-variable
time-driven control is achieved using a *sampled-data controller,* usually referred
to as a *digital controller.* In digital control of a process with input $c(t)$ and output $y(t)$, the error signal $e(t) = r(t) - y(t)$ is first sampled, which results in the
sampled error signal $e(nT)$ where T is the sampling interval and $n = \pm 0, \pm 1, \pm 2,$
... is the integer-valued discrete-time index. A digital controller can be viewed
as a discrete-time system that processes the discrete-time input $e(nT)$ to produce
a discrete-time control signal $c(nT)$. The control input $c(t)$ applied to the process
is generated from the discrete-time control signal $c(nT)$ by using a hold operation. The overall closed-loop configuration with digital controller is illustrated
in Figure 5.1.

To achieve good performance when using digital control, the sampling interval T must be selected to be sufficiently small. This is equivalent to choosing
the sampling frequency $f_s = 1/T$ Hz (cycles/sec) to be sufficiently fast. A common
rule of thumb is that f_s should be at least 20 times the bandwidth of the overall
closed-loop system, where the bandwidth refers to the frequency response of the
closed-loop system. The bandwidth can be shown to be approximately equal to
$1/(2\pi\tau)$ Hz, where τ is the time constant of the closed-loop system; that is, τ is
the amount of time it takes for the step response to reach 63% of the steady-state
value.

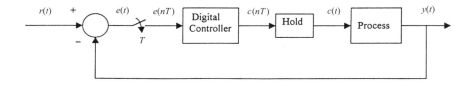

Figure 5.1   Closed-loop configuration with digital controller.

Then setting

$$f_s = \frac{1}{T} = 20 \left[\frac{1}{2\pi\tau} \right]$$

and solving for the sampling time T yields

$$T = \frac{\pi\tau}{10}$$

This can be taken as a maximum value of T for digital control.

    A digital controller can be implemented using a digital signal processing (DSP) chip, microprocessor, or PC. The implementation includes an analog-to-digital (A/D) conversion operation that samples, quantizes, and encodes the analog signal input to produce a binary signal (digital signal) consisting of 0s and 1s. A DSP chip, microprocessor, or PC then implements a set of operations (usually given by a program) for processing the digital signal to produce a digital control signal. In the last step of the implementation, the digital output signal is converted to an analog control signal using a digital-to-analog (D/A) converter. In a PC implementation, the A/D and D/A converters are on cards that plug into the computer, and the PC is programmed to realize the desired digital control action. Various software packages such as LABVIEW are available for implementing digital control on a PC.

    The design of a digital controller involves the use of the z transform, which is introduced next.

The z Transform

Given a sampled (discrete-time) signal $f(nT)$, where T is the sampling interval, the z transform of $f(nT)$ is a function $F(z)$ of the complex variable z. The z transforms of some common discrete-time signals are as follows:

$$f(nT) = c_o, \quad n = 0, \quad f(nT) = 0, \quad n \neq 0 \rightarrow F(z) = c_o$$

$$f(nT) = c_1, \quad n = 1, \quad f(nT) = 0, \quad n \neq 1 \rightarrow F(z) = c_1 z^{-1}$$

$$f(nT) = c, \quad n \geq 0 \rightarrow F(z) = \frac{cz}{z-1}$$

$$f(nT) = c(a^n), \quad n \geq 0 \rightarrow F(z) = \frac{cz}{z-a}$$

$$f(nT) = b \cos(\Omega n), \quad n \geq 0 \rightarrow F(z) = \frac{z^2 - (\cos\ \Omega)z}{z^2 - (2\cos\ \Omega)z + 1}$$

$$f(nT) = b \sin(\Omega n), \quad n \geq 0 \rightarrow F(z) = \frac{(\sin\ \Omega)z}{z^2 - (2\cos\ \Omega)z + 1}$$

Note that for these sampled signals, the z transform is independent of the sampling interval T. This turns out to be true in general; that is, the z transform of a sampled signal is independent of the sampling interval. In a given application, the sampling interval T is fixed, and it is understood that all discrete-time signals are defined relative to T. Thus there is no need to make explicit mention of T.

For each of the sampled signals $f(nT)$ with z transform $F(z)$ given above, the signal $f(nT)$ is equal to the inverse z transform of $F(z)$. A large collection of z transforms can be generated from the above transforms by using linearity of the z transform. Linearity means that if $F(z)$ is the z transform of $f(nT)$ and $G(z)$ is the z transform of $g(nT)$, then for any scalars a and b, the z transform of $af(nT) + bg(nT)$ is equal to $aF(z) + bG(z)$.

Now consider a system that processes a sampled signal input $c(nT)$ to produce a sampled signal output $y(nT)$. Such a system is said to be a discrete-time system since it processes a discrete-time input signal to produce a discrete-time output signal. The system is said to be linear if the response to $ac_1(nT) + bc_2(nT)$ is equal to $ay_1(nT) + by_2(nT)$, where $y_1(nT)$ and $y_2(nT)$ are the responses to $c_1(nT)$ and $c_2(nT)$, respectively. Here we are assuming that the system is at rest before the application of an input, where "at rest" means that $y(nT) = 0$ for $n = -1, -2, \ldots$.

A discrete-time system is said to be time invariant or constant if for any integer q, the response to the time-shifted input $c(nT-qT)$ is equal to the time-shifted output $y(nT-qT)$, where $y(nT)$ is the response to $c(nT)$. We are again assuming that the system is at rest before the application of an input.

If a discrete-time system is both linear and time invariant, it can be described in terms of its transfer function representation given by

$$Y(z) = G(z)C(z) \tag{5.1}$$

where $Y(z)$ is the z transform of the output $y(nT)$, $C(z)$ is the z transform of the input $c(nT)$, and $G(z)$ is the transfer function. In (5.1) it is assumed that the input $c(nT)$ is applied for $n = 0, 1, 2, \ldots$, and the system is at rest before the application of $c(nT)$.

To illustrate the construction of the transfer function representation (5.1), consider the discrete-time system given by the input/output equation

$$y(nT+T) = ay(nT) + bc(nT) \tag{5.2}$$

where a and b are constants. Equation (5.2) is an example of a *first-order difference equation*. The response $y(nT)$ to an input $c(nT)$ is computed from (5.2) by first setting $n = 0$ and $n = 1$. This gives

$$y(T) = ay(0) + bc(0) \tag{5.3}$$

$$y(2T) = ay(T) + bc(T) \tag{5.4}$$

Using (5.3) in (5.4) yields

$$y(2T) = a[ay(0) + bc(0)] + bc(T)$$

$$y(2T) = a^2 y(0) + abc(0) + bc(T) \tag{5.5}$$

This process, which is called *iteration*, is continued by setting $n = 2$ in (5.2) and using (5.5), which results in an expression for $y(3T)$.

To determine the transfer function representation for the discrete-time system given by (5.2), we apply the z transform to both sides of (5.2). This gives

$$zY(z) - y(0)z = aY(z) + bC(z) \tag{5.6}$$

where we are using the property that the z transform of the time-shifted signal $y(nT+T)$ is equal to $zY(z) - y(0)z$. We assume that $c(nT) = 0$ for $n < 0$ and the system is at rest before the application of $c(nT)$ so that $y(-T) = 0$. Then setting $n = -1$ in (5.2), we have

$$y(0) = ay(-T) + bc(-T) = 0 + 0 = 0$$

Hence (5.6) reduces to

$$zY(z) = aY(z) + bC(z)$$

and solving for $Y(z)$ gives the transfer function representation

$$Y(z) = \frac{b}{z-a} C(z)$$

In this case the transfer function is

$$G(z) = \frac{b}{z-a}$$

This transfer function is said to be first order since the degree of the denominator is equal to one.

The generalization to second-order and higher-order difference equations is straightforward and is therefore omitted (see Appendix A for an appropriate reference).

Design Of Digital Controllers

One approach to the design of a digital controller is first to design an analog controller given by transfer function $G_c(s)$, and then construct a digital controller by discretizing-in-time the analog controller to generate the transfer function $G_c(z)$ for the digital controller. Unfortunately, $G_c(z)$ is not simply equal to $G_c(s)$ with s replaced by z. (Life is never that simple!) We illustrate the discretization process by considering a PI controller.

Digital PI Controller

Recall from the previous chapter that the PI controller is given by

$$c(t) \;=\; K_p e(t) + K_I \int_{-\infty}^{t} e(\lambda)\, d\lambda \qquad\qquad (5.7)$$

where $c(t)$ is the control input to the process, $e(t) = r(t) - y(t)$ is the tracking error, and K_P and K_I are the controller gains. Then setting $t = nT+T$ in (5.7), where T is the sampling interval, yields

$$c(nT + T) \;=\; K_p e(nT + T) + K_I \int_{-\infty}^{nT+T} e(\lambda)\, d\lambda \qquad\qquad (5.8)$$

The second term on the right-hand side of (5.8) can be expressed in the form

$$K_I \int_{-\infty}^{nT+T} e(\lambda)\, d\lambda = K_I \int_{-\infty}^{nT} e(\lambda)\, d\lambda \;+\; K_I \int_{nT}^{nT+T} e(\lambda)\, d\lambda \qquad\qquad (5.9)$$

But from (5.7) we have

$$K_I \int_{-\infty}^{nT} e(\lambda)\, d\lambda = c(nT) - K_p e(nT) \qquad\qquad (5.10)$$

and we can approximate the second term on the right-hand side of (5.9) by

$$K_I \int_{nT}^{nT+T} e(\lambda)\, d\lambda = K_I T e(nT) \tag{5.11}$$

Inserting (5.10) and (5.11) into (5.9) gives

$$K_I \int_{-\infty}^{nT+T} e(\lambda)\, d\lambda = c(nT) - K_p e(nT) + K_I T e(nT)$$

and using this in (5.8) yields

$$c(nT+T) = K_p e(nT+T) + c(nT) - K_p e(nT) + K_I T e(nT)$$

Rearranging terms, we have the following first-order difference equation describing the discretized PI controller

$$c(nT+T) = c(nT) + K_p[e(nT+T) - e(nT)] + K_I T e(nT) \tag{5.12}$$

We refer to the controller given by (5.12) as the *digital PI controller*. The difference equation (5.12) can be programmed on a PC in order to implement the control action. This requires that the correct initial condition $c(0)$ be used. To determine $c(0)$, first set $n = -1$ in (5.12), which gives

$$c(0) = c(-T) + K_p[e(0) - e(-T)] + K_I T e(-T) \tag{5.13}$$

We are assuming that $c(-T) = y(-T) = r(-T) = 0$ and, thus,

$$e(-T) = r(-T) - y(-T) = 0$$

and from (5.13), we have

$$c(0) = K_p e(0) = K_p[r(0) - y(0)] \tag{5.14}$$

The digital PI controller given by (5.12) can be described in terms of its transfer function representation, which can be determined by taking the z transform of both sides of (5.12). This gives

$$zC(z) - c(0)z = C(z) + K_P[zE(z) - e(0)z - E(z)] + K_I TE(z) \quad (5.15)$$

From (5.14), $c(0) = K_P e(0)$, and thus (5.15) reduces to

$$zC(z) = C(z) + K_P[zE(z) - E(z)] + K_I TE(z)$$

Then, solving for $C(z)$, we have

$$C(z) = \frac{K_P z - K_P + K_I T}{z - 1} E(z) \quad (5.16)$$

From (5.16) we see that the transfer function $G_c(z)$ of the digital PI controller is

$$G_c(z) = \frac{K_P z - K_P + K_I T}{z - 1} \quad (5.17)$$

Note that $G_c(z)$ has a pole at $z = 1$ and a zero at $z = (K_P - K_I T)/K_P$. The pole at $z = 1$ is the "discrete-time counterpart" of the pole at $s = 0$ in the analog PI controller given by the transfer function $G_c(s) = K_P + K_I/s$. In other words, in the discretization process the pole at $s = 0$ in $G_c(s)$ is "mapped" to the pole at $z = 1$ in $G_c(z)$.

The transfer function $G_c(z)$ of the digital PI controller given by (5.17) can be expressed in the form

$$G_c(z) = \frac{K_P(z - 1) + K_I T}{z - 1}$$

$$G_c(z) = K_P + \frac{K_I T}{z - 1} \quad (5.18)$$

Equation (5.18) reveals the general form of the PI controller. Namely, from (5.18) we see that it includes proportional control with gain K_P and it contains a term that is the discrete-time counterpart to integral control.

Digital Control of a Motor

To illustrate the procedure for designing a digital PI controller, we again consider velocity control of a motor shaft, with the transfer function representation of the motor given by

$$\dot{\Theta}(s) = \frac{5}{s + 0.1} V(s)$$

where $\dot{\Theta}(s)$ is the LT of the angular velocity $\dot{\theta}(t)$ of the motor shaft and $V(s)$ is the LT of the voltage applied to the field circuit of the motor. In Chapter 4 we designed an analog PI controller for tracking a desired angular velocity, with the transfer function $G_c(s)$ of the controller given by

$$G_c(s) = 0.28 + \frac{0.1}{s}$$

so that $K_P = 0.28$ and $K_I = 0.1$.

With this controller, the time constant of the closed-loop system's step response was found to be approximately 0.7 sec (see Figure 4.4). Hence, by the criterion for the sampling interval T given previously, the maximum value of T that should be used is

$$T = \frac{\pi\tau}{10} = 0.22 \text{ second}$$

We select T to be 0.2 sec. Then from (5.18), we have that the transfer function $G_c(z)$ for a digital control implementation of the controller is given by

$$G_c(z) = 0.28 + \frac{(0.1)(0.2)}{z - 1} = 0.28 + \frac{0.02}{z - 1}$$

To determine the performance of the closed-loop system with this controller, we can compute the sampled output $\dot{\theta}(nT)$ as follows.

First, we assume that the hold operation in the feedback control configuration (see Figure 5.1) is a zero-order hold. This means that the output $c(t)$ of the hold is given by

$$c(t) = c(nT), \quad nT < t \le nT + T \tag{5.19}$$

Since the input to the hold operation is the discrete-time control signal $c(nT)$, from (5.19) we see that when the zero-order hold receives an input value $c(nT)$ at time $t = nT$, it holds this value until the next value of the input is received at time $t = nT + T$. For example, suppose that $c(nT)$ is the sampled version of 1 $- e^{-t}$ for $t \ge 0$ with $T = 0.5$ sec. Then the output of the zero-order hold is shown in Figure 5.2.

When the hold operation is a zero-order hold, the z transform $Y(z)$ of the sampled output $y(nT)$ is given by

$$Y(z) = G_p(z)C(z)$$

where $C(z)$ is the z transform of the discrete-time control signal $c(nT)$ and $G_p(z)$ is the transfer function of the discretization-in-time of the process based on the zero-order hold operation. The transfer function $G_p(z)$ can be computed directly from the transfer function $G_p(s)$ of the process by using the MATLAB command

$$gd = c2d(g, T, 'zoh')$$

where g is the given process transfer function, gd is the "discrete transfer function," T is the sampling interval, and 'zoh' stands for "zero-order hold."

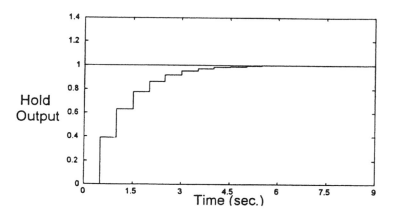

Figure 5.2   Output of zero-order hold.

For the motor example given earlier, we have

$$g = tf(5,[1 \ .1])$$

$$gd = c2d(g,0.2,\text{'zoh'})$$

Executing this on MATLAB results in the discrete transfer function

$$G_p(z) = \frac{0.9901}{z - 0.9802}$$

Now with the discretization of the controller given by $C(z) = G_c(z)E(z)$ and the discretization of the plant given by $Y(z) = G_p(z)C(z)$, we can compute the discrete closed-loop transfer function representation as follows: First, we have

$$C(z) = G_c(z)E(z) = G_c(z)[R(z) - Y(z)]$$

where $R(z)$ is the z transform of the sampled reference $r(nT)$. Then

$$Y(z) = G_p(z)C(z) = G_p(z)G_c(z)[R(z) - Y(z)]$$

and solving for $Y(z)$ gives the discrete closed-loop transfer function representation

$$Y(z) = \frac{G_p(z)G_c(z)}{1 + G_p(z)G_c(z)} R(z) \qquad (5.20)$$

From (5.20) we see that the closed-loop transfer function $G_{cl}(z)$ (in the z domain) is given by

$$G_{cl}(z) = \frac{G_p(z)G_c(z)}{1 + G_p(z)G_c(z)} \qquad (5.21)$$

We can then compute the sampled output $y(nT)$ by using (5.20). The proce-dure is illustrated here.

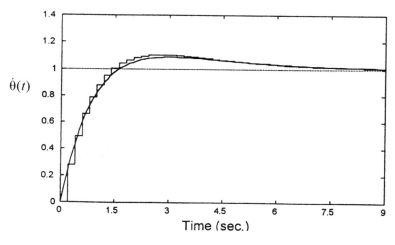

Figure 5.3 Unit-step response of closed-loop system with analog and
digital PI controllers.

For the motor example with

$$G_c(z) = 0.28 + \frac{0.02}{z-1} \quad \text{and} \quad G_p(z) = \frac{0.9901}{z - 0.9802}$$

the discrete closed-loop transfer function is

$$G_{cl}(z) = \frac{0.27723z - 0.25743}{z^2 - 1.70297z + 0.72277}$$

The zero-order hold of the discrete-time unit-step response of the closed-loop
system can be plotted using the MATLAB commands

$$\text{gd} = \text{tf}([0.27723 - 0.25743],[1 - 1.70297\ 0.72277],0.2)$$
$$\text{step(gd)}$$

where 0.2 in the specification for gd is the sampling interval. Running these
commands results in the "staircase plot" shown in Figure 5.3. Note that the
response values are held from sample time to sample time. Also plotted in Figure 5.3 is the step response of the closed-loop system with the analog PI controller (see Figure 4.4). It is clear from the figure that the performance of the

closed-loop system with the digital PI controller is very close to the performance with the analog PI controller.

An Alternate Design Approach

Instead of discretizing a given analog controller, we can first discretize the process transfer function $G_p(s)$ to generate a discrete process transfer function $G_p(z)$, and then we can design the digital controller transfer function $G_c(z)$ using $G_p(z)$. As noted before, $G_p(z)$ can be computed from $G_p(s)$ using the MATLAB command c2d.

The design of the digital controller transfer function $G_c(z)$ is based on the discrete-time closed-loop configuration shown in Figure 5.4. The closed-loop transfer function $G_{cl}(z)$ is the discrete transfer function given by (5.21). In correspondence with the continuous-time (analog) framework, the values of z for which $G_{cl}(z) = \infty$ are the poles of $G_{cl}(z)$, or the poles of the discrete-time closed-loop system.

To track a step input $r(nT) = r_o$, $n \geq 0$, the following requirements must be satisfied:

1. To achieve zero steady-state error, it is necessary for $G_p(z)G_c(z)$ to have a pole at $z = 1$. To achieve zero steady-state error when there is a step disturbance, $G_c(z)$ must have a pole at $z = 1$. These results are the discrete-time counterpart to the requirement that there be a pole at $s = 0$ in the analog case.

2. The magnitudes of all the closed-loop poles must be strictly less than 1. This is equivalent to requiring that all closed-loop poles be located within the unit disk of the complex plane, where the unit disk consists of all those complex numbers z such that $|z| < 1$. The unit disk is the region in the complex plane shown by the shaded area in Figure 5.5. The discrete-time closed-loop system is stable when all the closed-loop poles are located within the unit disk.

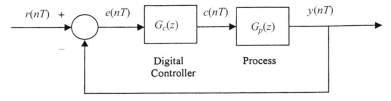

Figure 5.4 Discrete-time closed-loop control configuration

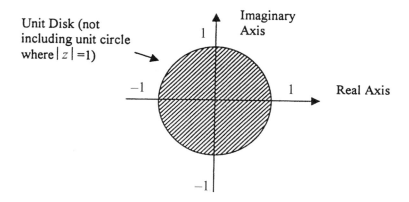

Figure 5.5 Unit disk.

3. For the transient part of the output response to be suitably fast, the closed-loop poles must be sufficiently far within the unit disk (i.e., sufficiently close to the origin). If all the closed-loop poles are placed at $z = 0$, the output response will be exactly zero after a finite number of sample times. This is called *dead-beat control*.

Note that the first requirement is satisfied if we use a digital PI controller with transfer function

$$G_c(z) = \frac{Az + B}{z - 1} \qquad (5.22)$$

To achieve tracking with this controller, we must be able to select the controller gains A and B so that the closed-loop poles are within the unit disk.

To illustrate the design process, we again consider digital control of the velocity of the motor with sampling interval $T = 0.2$ sec and with the discrete transfer function of the motor given by

$$G_p(z) = \frac{0.9901}{z - 0.9802}$$

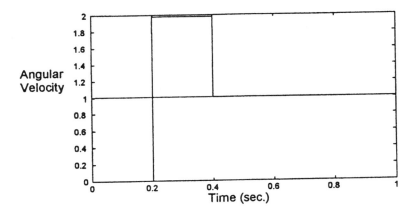

Figure 5.6 Unit-step response with dead-beat control.

With $G_c(z)$ given by (5.22), the closed-loop transfer function $G_{cl}(z)$ is

$$G_{cl}(z) = \frac{0.9901Az + 0.9901B}{z^2 + (0.9901A - 1.9802)z + 0.9901B + 0.9802} \qquad (5.23)$$

From (5.23), we see that the two closed poles can be put anywhere we desire within the unit disk. We shall achieve dead-beat control by putting both poles at $z = 0$. This will be the case when

$$0.9901A - 1.9802 = 0 \quad \text{and} \quad 0.9901B + 0.9802 = 0$$

Solving for A and B gives

$$A = 2 \quad \text{and} \quad B = -0.99$$

and

$$G_{cl}(z) = \frac{1.982z - 0.980}{z^2}$$

We can then use the MATLAB commands

$$gd = tf([1.982 -0.98],[1\ 0\ 0], 0.2)$$
$$step(gd)$$

to plot the zero-order hold of the unit-step response of the discrete-time closed-loop system. The result is shown in Figure 5.6. Note that the output is exactly equal to the reference input after two sample times, but there is a substantial amount of overshoot. To eliminate the overshoot we can use a modified digital PI controller (see Problem 5.6).

Problems

5.1   Consider the discrete-time system with input $c(n)$ and output $y(n)$ given by

$$y(n+1) = -0.5y(n) + c(n)$$

(a)   Compute $y(n)$ for $n = 0, 1, 2, 3, 4$ when $y(-1) = 1$ and $c(n) = 0$ for $n < 0$, $c(n) = 1$ for $n \geq 0$.

(b)   Determine the transfer function $G_p(z)$ of the system.

(c)   When $y(-1) = 0$ and $c(n) = 0$ for $n < 0$, $c(n) = 1$ for $n \geq 0$, compute the z transform $Y(z)$ of the response $y(n)$.

(d)   Using your result from part (c), derive an expression for $y(n)$ for all $n \geq 0$.

(e)   Repeat parts (a),(b),(c), and (d) with $y(-1) = 1$.

5.2   Consider the analog (continuous-time) system given by the transfer function

$$G_p(s) = \frac{5}{s + 10}$$

(a)   Design an analog PI controller so that the unit-step response of the closed-loop system is equal to

$$y(t) = 1 + c_1 e^{-2t} + c_2 e^{-5t}, \quad t \geq 0$$

where c_1 and c_2 are constants.

(b) Using the method given in this chapter, discretize the PI controller found in part (a). Express your answer by giving the transfer function $G_c(z)$ of the resulting digital controller.

(c) Discretize the analog process with transfer function $G_p(s)$ using the zero-hold operation and compute the transfer function of the resulting discrete-time closed-loop system with the digital PI controller.

(d) Use MATLAB to plot the zero-order hold of the unit-step response of the discrete-time closed-loop system obtained in part (c). On the same graph plot the unit-step response of the analog system with the PI controller designed in part (a). Compare the results. What do you conclude?

5.3 Consider the analog (continuous-time) system given by the transfer function

$$G_p(s) = \frac{2}{s+5}$$

(a) By first designing an analog PI controller, design a digital PI controller so that the sampled output response $y(nT)$ of the closed-loop system resulting from the reference input $r(nT) = r_0$, $n \geq 0$, is

$$y(nT) = r_0 + c_1 \left(e^{-2T} \right)^n + c_2 \left(e^{-4T} \right)^n, \quad n \geq 0$$

where c_1 and c_2 are constants and $T = 0.2$. Express your answer by giving the transfer function $G_c(z)$ of the resulting digital controller with all coefficients evaluated.

(b) By first discretizing the given process using the hold operation, design a digital PI controller so that the sampled output response $y(nT)$ of the closed-loop system resulting from the reference input $r(nT) = r_0$, $n \geq 0$, is

$$y(nT) = r_0 + c_1 \left(e^{-2T} \right)^n + c_2 \left(e^{-4T} \right)^n, \quad n \geq 0$$

where c_1 and c_2 are constants and $T = 0.2$. Express your answer by giving the transfer function $G_c(z)$ of the resulting digital controller with all coefficients evaluated.

(c) Would you expect the digital PI controllers constructed in parts (a) and (b) to be the same? Explain why or why not.

5.4 An analog system with input $c(t)$ and output $y(t)$ has transfer function $G_p(s) = 1/(s+1)$. The system is to be controlled using a digital controller given by

$$C(z) = R(z) + \left[A + \frac{B}{z-1} \right] \left[R(z) - Y(z) \right]$$

where $C(z)$ is the z transform of $c(t)$, $R(z)$ is the z transform of the reference $r(t)$, and $Y(z)$ is the z transform of $y(t)$. Find the values of A and B so that the poles of the discretized closed-loop system are equal to $-0.9, -0.9$.

5.5 Consider the tank given by $dy(t)/dt = 0.5iv(t)$, where $y(t)$ is the level in inches of liquid in the tank and $iv(t)$ is the position of the input valve. Recall that $0 \le iv(t) \le 1$.

(a) Design a modified analog PI controller so that the closed-loop system has two poles equal to $-0.05, -0.05$. Express your answer by giving the gains K_P, K_I, and K_F.

(b) Discretize the modified PI controller found in part (a) with $T = 1$. Express your answer by giving the equation for the control input $iv(n)$ to the tank.

(c) By using the discretization of the tank based on the zero-order hold, plot the zero-order hold of the response of the closed-loop system with the controller in part (b) and with the reference input $r(t) = 10$ in for $t \ge 0$.

5.6 In this chapter a digital PI controller was designed for velocity control of the motor with the discrete transfer function

$$G_p(z) = \frac{0.9901}{z - 0.9802}$$

(a) Using this discrete transfer function, design a modified digital PI controller so that the zero of the closed-loop system due to the controller is equal to 0 and it cancels one of the closed-loop poles with the other closed-loop pole equal to 0. Express your answer by giving the difference equation for the controller.

(b) Using MATLAB, plot the zero-order hold of the unit-step response of the closed-loop system with the modified digital PI controller in part (a)

5.7 Consider the tank in Problem 5.5.

(a) By first discretizing the tank using the zero-order hold operation, design a modified digital PI controller so that $y(n) \to 10$ inches as $n \to \infty$ and the poles of the discretized closed-loop system are equal to 0.96, 0.96. Express your answer by giving the difference equation for the digital controller.

(b) Using MATLAB, plot the zero-order hold of the output $y(t)$ and the control input $iv(n)$ when $r(t) = 10$ in for $n \geq 0$. Is it true that $0 \leq iv(t) \leq 1$? If not, there is an error in your design.

5.8 Determine the transfer function of a digital PID controller by discretizing the analog PID controller given by $G_c(s) = K_P + K_I/s + K_D s$. Use the discretization method given this chapter . Express your answer by giving the transfer function $G_c(z)$ of the resulting digital controller.

5.9 Consider the analog (continuous-time) system given by the transfer function

$$G_p(s) = \left[\frac{1}{s+1}\right]\left[\frac{1}{s-1}\right]$$

(a) Design a digital controller so that the sampled output of the closed-loop system's unit-step response is

$$y(nT) = 1 + c_1\left(e^{-1}\right)^n + c_2\left(e^{-2}\right)^n \cos(n + \theta), \quad n \geq 0$$

where c_1, c_2, and θ are constants.

(b) Discretize the analog process with transfer function $G_p(s)$ using the zero-hold operation and compute the transfer function of the resulting discrete-time closed-loop system with the digital PI controller.

(c) Use MATLAB to plot the zero-order hold of the unit-step response of the discrete-time closed-loop system obtained in part (b).

5.10 Repeat Problem 5.9 for the analog process with the transfer function

$$G_p(s) = \frac{1}{s^2 + 2s - 1}$$

Chapter 6

Model Predictive, Adaptive, and Neural Net Controllers

In the previous chapters, the continuous-variable time-driven controllers that were considered were limited to proportional (P), proportional-plus-integral (PI), and proportional-plus-integral-plus-derivative (PID) controllers. Although these types of controllers are known to work well in a wide range of applications, in practice, many systems cannot be effectively controlled using P, PI, or PID controllers. In these cases more complicated controllers are necessary to achieve the desired performance. In this chapter we give a brief introduction to more general types of controllers, beginning with model predictive controllers.

Model Predictive Controllers

A very powerful control methodology is *model predictive control* (MPC), which was first developed for the chemical process industries and is now being applied to a rapidly expanding range of technologies. The basic idea of MPC is to use a model to predict the output response of a system at various time points in the future, and based on this, control inputs are computed to yield the desired response. To illustrate the computation of the control in one type of MPC, we begin by considering a system with the sampled input–output representation given by

$$y(nT+T) = ay(nT) + bc(nT) \tag{6.1}$$

In (6.1), T is the sampling interval, n is the discrete-time index (n = 0, 1, 2, ...), $y(nT)$ and $c(nT)$ are the system output and control input at time t = nT, and a and b are fixed constants. To simplify the notation, throughout this chapter we set T = 1, so that (6.1) reduces to

$$y(n+1) = ay(n) + bc(n) \tag{6.2}$$

Now suppose that the control objective is to have the output $y(n)$ track a given reference $r(n)$; that is, we want $y(n) = r(n)$. The "predictive model" for achieving this control objective is given by (6.2); in other words, using (6.2) we can predict the value $y(n+1)$ of the output at time $n+1$ using the value $y(n)$ of the output at time n and the value $c(n)$ of the input at time n. Since we want $y(n) = r(n)$, to find the control that will produce this result we simply set $y(n+1) = r(n+1)$ and solve (6.2) for $c(n)$. This yields

$$c(n) = -\frac{a}{b}y(n) + \frac{1}{b}r(n+1) \tag{6.3}$$

From (6.3), we see that the control input $c(n)$ at time n is computed from the output $y(n)$ at time n and the reference $r(n+1)$ at time $n+1$. To verify that the control given by (6.3) works, simply insert (6.3) into (6.2). This gives $y(n+1) = r(n+1)$ for $n \geq 0$ where the initial condition is at time n = 0.

To consider a second example, we modify the system model (6.2) so that

$$y(n+1) = ay(n) + bc(n-1) \tag{6.4}$$

In this case, setting $y(n+1) = r(n+1)$ in (6.4) and solving for $c(n-1)$ gives

$$c(n-1) = -\frac{a}{b}y(n) + \frac{1}{b}r(n+1) \tag{6.5}$$

However, the control given by (6.5) cannot be implemented since the control $c(n-1)$ at time $n-1$ cannot depend on the output $y(n)$ at future time n.

To generate a control for the system given by (6.4), we can use the prediction of the output at time $n+2$: First, replacing n by $n+1$ in (6.4) gives

$$y(n+2) = ay(n+1) + bc(n) \tag{6.6}$$

and using (6.4), we have

$$y(n+2) = a[ay(n) + bc(n-1)] + bc(n)$$

$$y(n+2) = a^2 y(n) + abc(n-1) + bc(n) \qquad (6.6)$$

Then setting $y(n+2) = r(n+2)$ in (6.6) and solving for $c(n)$ results in the control

$$c(n) = -\frac{a^2}{b} y(n) - ac(n-1) + \frac{1}{b} r(n+2) \qquad (6.7)$$

With the control given by (6.7), we have $y(n+2) = r(n+2)$ for $n \geq 0$ starting with the initial conditions at time $n = -1$ and 0.

General Case

The control methodology given above easily generalizes to a system given by

$$y(n+1) = \sum_{i=0}^{p} a_i y(n-i) + b_0 c(n) + \sum_{i=1}^{q} b_i c(n-i) \qquad (6.8)$$

where the a_i and the b_i are fixed constants. To generate a control that will result in $y(n)$ tracking a reference $r(n)$, we set $y(n+1) = r(n+1)$ in (6.8) and solve for $c(n)$. This yields

$$c(n) = -\sum_{i=0}^{p} \frac{a_i}{b_0} y(n-i) - \sum_{i=1}^{q} \frac{b_i}{b_0} c(n-i) + \frac{1}{b_0} r(n+1) \qquad (6.9)$$

If $b_0 = 0$, we need to use the prediction of the output at time $n+2$: Replacing n by $n+1$ in (6.8) gives

$$y(n+2) = a_0 y(n+1) + \sum_{i=1}^{p} a_i y(n+1-i) + b_1 c(n) + \sum_{i=2}^{q} b_i c(n+1-i) \qquad (6.10)$$

Inserting (6.8) into (6.10), setting $y(n+2) = r(n+2)$, and solving (6.10) for $c(n+1)$ results in the desired control. We do not attempt to write down the expression for the control. If $b_1 = 0$, we can use the prediction of the output at time $n+3$, and so on. The details are omitted.

Robustness

The control resulting from this methodology may not perform well if there are disturbances or there is model error [i.e., the model given by (6.8) is not accurate]. In particular, if there is a constant disturbance, there will be a steady-state tracking error. To show this, we first modify the model (6.8) by adding a disturbance term $w(n)$:

$$y(n+1) = \sum_{i=0}^{p} a_i y(n-i) + b_0 c(n) + \sum_{i=1}^{q} b_i c(n-i) + w(n) \qquad (6.11)$$

Inserting (6.9) into (6.11) gives

$$y(n+1) = r(n+1) + w(n) \qquad (6.12)$$

and thus the disturbance $w(n)$ directly affects the tracking performance. If $w(n)$ is equal to a constant w_o, then from (6.12) we see that there will be a steady-state tracking error equal to w_o.

Disturbance Estimator

The effect of a disturbance $w(n)$ can be reduced by attempting to estimate the disturbance and including the estimate in the control given by (6.9). We illustrate the process of estimating the disturbance by considering the first-order system:

$$y(n+1) = ay(n) + bc(n) + w(n) \qquad (6.13)$$

where $w(n)$ is the disturbance. We take the estimate $\widehat{w}(n)$ of the disturbance at time n to be

$$\widehat{w}(n) = y(n) - ay(n-1) - bc(n-1) \qquad (6.14)$$

If $w(n)$ is constant [$w(n) = w$ for all n], it is easy to see that $\hat{w}(n) = w(n)$ for all $n \geq 1$ (see Problem 6.4). If $w(n)$ is slowing varying as a function of n, the estimate $\hat{w}(n)$ will be "close to" $w(n)$.

With the estimate of $w(n)$ given by (6.14), we take the control input to be

$$c(n) = -\frac{a}{b}y(n) + \frac{1}{b}r(n+1) - \frac{1}{b}\hat{w}(n) \qquad (6.15)$$

where $r(n)$ is the reference input. Inserting (6.15) into (6.13) gives

$$y(n+1) = r(n+1) + w(n) - \hat{w}(n) \qquad (6.16)$$

From (6.16), we see that the tracking error is equal to the error in estimating $w(n)$. Therefore, the closer the estimate is to $w(n)$, the smaller the tracking error.

Optimal Control

In model predictive control, optimality constraints can be included in the solution for the control. For example, again consider the system given by

$$y(n+1) = ay(n) + bc(n) \qquad (6.17)$$

but now suppose that the objective is to find a control $c(n)$ that minimizes the criterion

$$[y(n+1) - r(n+1)]^2 + h[c(n)]^2 \qquad (6.18)$$

where $h \geq 0$. This is a type of optimal control problem where an optimal control is one that minimizes the sum of the squares in (6.18). The constant h in (6.18) is called a *weighting factor*. Note that when $h = 0$, we have the control problem considered previously. When $h > 0$, the magnitude $|c(n)|$ of the optimal control solution is constrained as a result of the term $h[c(n)]^2$. The larger h is, the stronger the constraint on the magnitude $|c(n)|$.

To compute a control that minimizes the sum of squares in (6.18), we first insert (6.17) into (6.18), which gives

$$[ay(n) + bc(n) - r(n+1)]^2 + h[c(n)]^2 \qquad (6.19)$$

We then take the partial derivative of (6.19) with respect to $c(n)$ and set the result to zero, which yields

$$2[ay(n) + bc(n) - r(n+1)]b + 2hc(n) = 0 \tag{6.20}$$

Finally, solving (6.20) for $c(n)$ yields the optimal control

$$c(n) = \frac{b}{b^2 + h}\left[-ay(n) + r(n+1)\right] \tag{6.21}$$

Note that when $h = 0$, the control given by (6.21) is the same as the control given by (6.3). The control given by (6.21) will result in a steady-state tracking error (see Problem 6.3). The tracking error is a consequence of the constraint on the magnitude of the control, so a price is paid in constraining the magnitude of the control.

MIMO Case

Unlike P, PI, or PID control, the MPC approach considered in this chapter extends to the case when the system to be controlled has multiple inputs and outputs. This is referred to as the multiple-input /multiple-output (MIMO) case. We illustrate the computation of the control by considering the following two-input/two-output system:

$$y_1(n + 1) = a_1 y_1(n) + a_2 y_2(n) + b_1 c_1(n) + b_2 c_2(n) \tag{6.22}$$

$$y_2(n + 1) = a_3 y_1(n) + a_4 y_2(n) + b_3 c_1(n) + b_4 c_2(n) \tag{6.23}$$

We can express these two equations in terms of a single vector equation as follows. First, we let $c(n)$ denote the two-element control vector defined by

$$c(n) = \begin{bmatrix} c_1(n) \\ c_2(n) \end{bmatrix}$$

and we let $y(n)$ be the two-element output vector defined by

$$y(n) = \begin{bmatrix} y_1(n) \\ y_2(n) \end{bmatrix}$$

Then in terms of the vectors $c(n)$ and $y(n)$, (6.22) and (6.23) can be written as a single vector equation given by

$$y(n+1) = Ay(n) + Bc(n) \qquad (6.24)$$

where A and B are the two-by-two matrices given by

$$A = \begin{bmatrix} a_1 & a_2 \\ a_3 & a_4 \end{bmatrix}, \quad B = \begin{bmatrix} b_1 & b_2 \\ b_3 & b_4 \end{bmatrix}$$

Now suppose we want $y_1(n)$ to track $r_1(n)$ and $y_2(n)$ to track $r_2(n)$ and we define

$$r(n) = \begin{bmatrix} r_1(n) \\ r_2(n) \end{bmatrix}$$

If the matrix B is invertible, we can set $y(n+1) = r(n+1)$ in (6.24) and solve for $c(n)$. This yields the control

$$c(n) = B^{-1}[r(n+1) - Ay(n)]$$

where B^{-1} is the inverse of B. The matrix B is invertible if and only

$$b_1 b_4 - b_2 b_3 \neq 0,$$

where $b_1 b_4 - b_2 b_3$ is the determinant of B. If B is not invertible, it will not be possible to find a control that forces $y_1(n)$ and $y_2(n)$ to track any given reference signals $r_1(n)$ and $r_2(n)$.

Now suppose the system is given by

$$y(n+1) = Ay(n) + Bc(n-1) \qquad (6.25)$$

In this case, we can compute a control that forces $y(n) = r(n)$ by using the prediction of the output vector at time $n+2$: Replacing n by $n+1$ in (6.25) gives

$$y(n+2) = Ay(n+1) + Bc(n)$$

and using (6.25) we have

$$y(n+2) = A[Ay(n) + Bc(n-1)] + Bc(n)$$

$$= A^2y(n) + ABc(n-1) + Bc(n) \qquad (6.26)$$

If B is invertible, setting $y(n+2) = r(n+2)$ in (6.26) and solving for $c(n)$ results in the control

$$c(n) = B^{-1}[r(n+2) - A^2y(n) - ABc(n-1)]$$

If B is not invertible, a solution (if there is one) to the tracking of a specific reference is more complicated and is left to a more advanced treatment of control.

Adaptive Control

In applications it is often the case that the coefficients of the system model are not known before the system is put into operation or the coefficients vary during system operation. For example, if the system is given by

$$y(n+1) = ay(n) + bc(n)$$

the coefficients a and b (also called the *system parameters*) may not be known a priori or they may vary during system operation. If a and b are varying so that they are functions $a(n)$ and $b(n)$ of n, then the model becomes

$$y(n+1) = a(n)y(n) + b(n)c(n) \qquad (6.27)$$

If $a(n)$ and $b(n)$ are known, then a control that forces $y(n)$ to be equal to a reference $r(n)$ is computed by setting $y(n+1) = r(n+1)$ in (6.27) and solving for $c(n)$. This results in the control

$$c(n) = -\frac{a(n)}{b(n)}y(n) + \frac{1}{b(n)}r(n+1) \qquad (6.28)$$

where it is assumed that $b(n) \neq 0$ for all values of n. However, if $a(n)$ and $b(n)$ are not known, it is necessary to estimate $a(n)$ and $b(n)$, and then the estimates may be used in the control given by (6.28). This results in a type of *adaptive control.* If $\hat{a}(n)$ and $\hat{b}(n)$ denote the estimates of $a(n)$ and $b(n)$ at time n, the adaptive control is given by

$$c(n) = -\frac{\hat{a}(n)}{\hat{b}(n)}y(n) + \frac{1}{\hat{b}(n)}r(n+1) \qquad (6.29)$$

where it is assumed that $\hat{b}(n) \neq 0$. If $\hat{b}(n) = 0$ when $n = n_o$, we can take $c(n_o) = c(n_o - 1)$.

There are various different approaches for estimating the coefficients $a(n)$ and $b(n)$ in (6.27). Here we consider a simple procedure that can work well if the "rate of change" of the coefficients $a(n)$ and $b(n)$ is not large: First, replacing n by $n-2$ in (6.27) gives

$$y(n-1) = a(n-2)y(n-2) + b(n-2)c(n-2) \qquad (6.30)$$

and replacing n by $n-1$ in (6.27) gives

$$y(n) = a(n-1)y(n-1) + b(n-1)c(n-1) \qquad (6.31)$$

If we assume that the value of $a(n-2)$ is close to the value of $a(n-1)$ and the value of $b(n-2)$ is close to the value of $b(n-1)$, we can express (6.30) and (6.31) in terms of the single vector equation

$$\begin{bmatrix} y(n-1) \\ y(n) \end{bmatrix} = \begin{bmatrix} y(n-2) & c(n-2) \\ y(n-1) & c(n-1) \end{bmatrix}\begin{bmatrix} a(n-1) \\ b(n-1) \end{bmatrix} \qquad (6.32)$$

If

$$y(n-2)c(n-1) - c(n-2)y(n-1) \neq 0 \qquad (6.33)$$

the matrix

$$\begin{bmatrix} y(n-2) & c(n-2) \\ y(n-1) & c(n-1) \end{bmatrix}$$

has an inverse, and thus in this case we can solve (6.32) for $a(n-1)$ and $b(n-1)$. The result is

$$
\begin{bmatrix} a(n-1) \\ b(n-1) \end{bmatrix} = \begin{bmatrix} y(n-2) & c(n-2) \\ y(n-1) & c(n-1) \end{bmatrix}^{-1} \begin{bmatrix} y(n-1) \\ y(n) \end{bmatrix} \tag{6.34}
$$

We can then define the estimates of $a(n)$ and $b(n)$ by $\hat{a}(n) = a(n-1)$ and $\hat{b}(n) = b(n-1)$ where $a(n-1)$ and $b(n-1)$ are given by (6.34). The estimates $\hat{a}(n)$ and $\hat{b}(n)$ are then used in (6.29) to generate the adaptive control.

It is important to note that the condition (6.33) is a type of *excitation condition*; that is, the input $c(n)$ (the excitation) and the output $y(n)$ must satisfy this condition in order for the given estimation procedure to work. There exist estimation schemes that work under a weaker condition than that given by (6.33), but these are not considered.

To illustrate the evaluation of (6.34), suppose that the input–output values of the system are $y(0) = 1$, $y(1) = 1.5$, $y(2) = 1.75$, $y(3) = 1.875$, $c(0) = 1$, $c(1) = 1$, $c(2) = 1$. Inserting these values into (6.34) with $n = 3$ yields

$$
\begin{bmatrix} a(2) \\ b(2) \end{bmatrix} = \begin{bmatrix} y(1) & c(1) \\ y(2) & c(2) \end{bmatrix}^{-1} \begin{bmatrix} y(2) \\ y(3) \end{bmatrix} = \begin{bmatrix} 1.5 & 1 \\ 1.75 & 1 \end{bmatrix}^{-1} \begin{bmatrix} 1.75 \\ 1.875 \end{bmatrix}
$$

$$
\begin{bmatrix} a(2) \\ b(2) \end{bmatrix} = \frac{1}{-0.25} \begin{bmatrix} 1 & -1 \\ -1.75 & 1.5 \end{bmatrix} \begin{bmatrix} 1.75 \\ 1.875 \end{bmatrix} = \begin{bmatrix} 0.5 \\ 1 \end{bmatrix}
$$

Thus, $\hat{a}(3) = 0.5$, $\hat{b}(3) = 1$, and using (6.29), we have that the value $c(3)$ of the control at time $n = 3$ is given by

$$
c(3) = -0.5y(3) + r(4)
$$

Direct Adaptive Control

The type of control just considered is referred to as *indirect adaptive control* since the coefficients or parameters of the system model are estimated first using input–output data, and then the estimates are used to generate the control. In contrast, in *direct adaptive control* the parameters of the control are directly determined from input-output data. We illustrate the construction of a direct adaptive controller by again considering the system given by

$$y(n+1) = a(n)y(n) + b(n)c(n) \tag{6.35}$$

If $b(n) \neq 0$, from (6.35) we see that $c(n)$ can be expressed as a function of $y(n+1)$ and $y(n)$; that is,

$$c(n) = -\frac{a(n)}{b(n)} y(n) + \frac{1}{b(n)} y(n+1)$$

Hence, we have

$$c(n) = \alpha(n)y(n) + \beta(n)y(n+1) \tag{6.36}$$

The basic idea in direct adaptive control is to directly estimate the coefficients $\alpha(n)$ and $\beta(n)$ in (6.36). Then with the estimates denoted by $\hat{\alpha}(n)$ and $\hat{\beta}(n)$, the direct adaptive control is given by

$$c(n) = \hat{\alpha}(n)y(n) + \hat{\beta}(n)r(n+1) \tag{6.37}$$

where $r(n)$ is the reference.

The estimates can be computed using a procedure similar to that given above for estimating the coefficients of the system model: Replacing n by $n-2$ and $n-1$ in (6.36) gives

$$c(n-2) = \alpha(n-2)y(n-2) + \beta(n-2)y(n-1) \tag{6.38}$$

$$c(n-1) = \alpha(n-1)y(n-1) + \beta(n-1)y(n) \tag{6.39}$$

Assuming that $\alpha(n-2) \approx \alpha(n-1)$ and $\beta(n-2) \approx \beta(n-1)$, we can write (6.38) and (6.39) in terms of the single vector equation

$$\begin{bmatrix} c(n-2) \\ c(n-1) \end{bmatrix} = \begin{bmatrix} y(n-2) & y(n-1) \\ y(n-1) & y(n) \end{bmatrix} \begin{bmatrix} \alpha(n-1) \\ \beta(n-1) \end{bmatrix} \tag{6.40}$$

If

$$y(n-2)y(n) - [y(n-1)]^2 \neq 0 \qquad (6.41)$$

the solution to (6.40) is

$$\begin{bmatrix} \alpha(n-1) \\ \beta(n-1) \end{bmatrix} = \begin{bmatrix} y(n-2) & y(n-1) \\ y(n-1) & y(n) \end{bmatrix}^{-1} \begin{bmatrix} c(n-2) \\ c(n-1) \end{bmatrix} \qquad (6.42)$$

We then take the estimates of $\alpha(n)$ and $\beta(n)$ to be $\hat{\alpha}(n) = \alpha(n-1)$ and $\hat{\beta}(n) = \beta(n-1)$. Note that the excitation condition (6.41) for direct adaptive control differs from the excitation condition (6.33) for indirect adaptive control.

To illustrate the evaluation of (6.42), again suppose that the system input-output values are $y(0) = 1$, $y(1) = 1.5$, $y(2) = 1.75$, $y(3) = 1.875$, $c(0) = 1$, $c(1) = 1$, $c(2) = 1$. Inserting these values into (6.42) when $n = 3$ yields

$$\begin{bmatrix} \alpha(2) \\ \beta(2) \end{bmatrix} = \begin{bmatrix} 1.5 & 1.75 \\ 1.75 & 1.875 \end{bmatrix}^{-1} \begin{bmatrix} 1 \\ 1 \end{bmatrix} = \frac{1}{-0.25} \begin{bmatrix} 1.875 & -1.75 \\ -1.75 & 1.5 \end{bmatrix} \begin{bmatrix} 1 \\ 1 \end{bmatrix}$$

$$\begin{bmatrix} \alpha(2) \\ \beta(2) \end{bmatrix} = \begin{bmatrix} -0.5 \\ 1 \end{bmatrix}$$

Hence, $\hat{\alpha}(3) = -0.5$, $\hat{\beta}(3) = 1$, and using (6.37), we have that the value $c(3)$ of the direct adaptive control at time $n = 3$ is given by

$$c(3) = -0.5y(3) + r(4)$$

Note that this value for $c(3)$ is the same as that obtained earlier using indirect adaptive control.

Neural Net Controllers

In many applications, the output $y(n+1)$ of a sampled system at time $n+1$ will be a nonlinear function of $y(n-i)$ and the control values $c(n-i)$ for some range of values of i. For example, the nonlinear version of the system given by

$$y(n+1) = ay(n) + bc(n) \qquad (6.43)$$

is the system given by

$$y(n+1) = f[y(n),c(n)] \qquad (6.44)$$

where $f[y(n),c(n)]$ is some nonlinear function of $y(n)$ and $c(n)$. The function $f[y(n),c(n)]$ is a linear function of $y(n)$ and $c(n)$ if and only if there exist constants a and b such that

$$f[y(n),c(n)] = ay(n) + bc(n) \qquad (6.45)$$

Hence, (6.44) reduces to (6.43) when $f[y(n),c(n)]$ is linear.

To generate a tracking control for the system given by (6.44), one can first attempt to identify the nonlinear function f by using a neural network. This leads to an *indirect neural net controller*. In contrast, in a *direct neural net controller*, one first attempts to express the control $c(n)$ as a function of $y(n)$ and $y(n+1)$; that is, using a neural network we attempt to identify the function g in the relationship

$$c(n) = g[y(n), y(n+1)]. \qquad (6.46)$$

The relationship given by (6.46) is a type of inverse of the relationship given by (6.44). In particular, if (6.46) is inserted into (6.44), the result will be $y(n+1) = y(n+1)$. Thus, if the function g is known and the control is taken to be

$$c(n) = g[y(n),r(n+1)] \qquad (6.47)$$

where $r(n)$ is the reference, the result will be $y(n+1) = r(n+1)$ as desired.

Given the system defined by (6.44), the inverse function g given by (6.46) may not exist. The inverse function g exists if the partial derivative of f with respect to $c(n)$ is nonzero for all values of n. Note that when f is linear so that it is given by (6.45), the partial derivative of f with respect to $c(n)$ is equal to b. Clearly, b must be nonzero in order to solve the control problem in the linear case.

Unless f is known, which is often not the case, the condition on the partial derivative of f cannot be checked, and thus one usually proceeds assuming that the inverse g does exist. A control that is designed based on the incorrect assumption that the inverse exists will not work. Testing the performance of the controlled system will discover this!

We now model the relationship (6.46) using a neural network that is defined in terms of *neurons*. A neuron with output v and inputs $u_1, u_2, \ldots, u_q$ is given by

$$v = act\left[\sum_{i=1}^{q} u_i\right]$$

where "act" is the activation function. Various choices are available for the activation function. One common example is the *arctan function* given by

$$act(\eta) = \frac{e^{\eta} - e^{-\eta}}{e^{\eta} + e^{-\eta}} \qquad (6.48)$$

A sketch of the arctan function is shown in Figure 6.1. Note that it "compresses" all input values into the range from -1 to 1.

We can model the relationship given by (6.46) using a two-neuron model, as shown in Figure 6.2. From the block diagram in the figure, we have

$$c(n) = d_1\chi_1(n) + d_2\chi_2(n) \qquad (6.49)$$

where

$$\chi_1(n) = act\left[a_{11}y(n) + b_{11}y(n+1)\right] \qquad (6.50)$$

$$\chi_2(n) = act\left[a_{12}y(n) + b_{12}y(n+1)\right] \qquad (6.51)$$

The constants $d_1, d_2, a_{11}, a_{12}, b_{11}, b_{12}$ in (6.49) – (6.51) are called the *weights* of the neural network. The signals $\chi_1(n)$ and $\chi_2(n)$ defined by (6.50) and (6.51) are located at the two *hidden nodes* of the neural network. These nodes are said to be "hidden" since they cannot be directly accessed from either the input nodes where $y(n)$ and $y(n+1)$ are applied or the output node where $c(n)$ is extracted. The neural network in Figure 6.2 is called a *feedforward neural network* with three layers consisting of the input layer, the hidden layer, and the output layer.

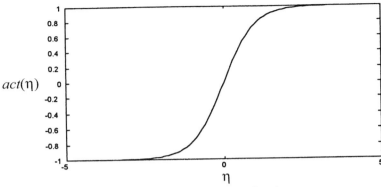

Figure 6.1 Plot of the arctan function

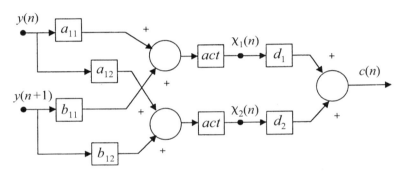

Figure 6.2 Block diagram of two-neuron neural network.

Any relationship of the form (6.46) can be modeled arbitrarily closely by a three-layer feedforward neural network if a sufficient number of nodes are used in the hidden layer. Unfortunately, there is no result that specifies how many hidden nodes must be used. This is determined by trial and error.

Now, to generate the neural net controller, we must determine the weights of the neural network given by (6.49) – (6.51). This is accomplished by training the neural network using input–output data. The training procedure is beyond the scope of this book and thus is not considered. See Appendix A for an appropriate reference.

Once the weights in (6.49) – (6.51) have been determined, the neural net controller is given by these equations with $y(n+1)$ replaced by $r(n+1)$; that is, the neural net controller is given by

$$c(n) = d_1\chi_1(n) + d_2\chi_2(n)$$

$$\chi_1(n) = act\Big[a_{11}y(n) + b_{11}r(n+1)\Big]$$

$$\chi_2(n) = act\Big[a_{12}y(n) + b_{12}r(n+1)\Big]$$

The weights in the neural net controller can be updated while the system is operating, and thus the neural net controller can handle variations in the weights due to variations in the system as it evolves in time.

Problems

6.1   A system with input $c(n)$ and output $y(n)$ is given by

$$y(n+1) + y(n) - y(n-1) = c(n-1)$$

(a) Using the MPC approach, design a controller so that $y(n)$ tracks a reference $r(n)$. Express your answer by giving the equation for the control $c(n)$.

(b) For your result in part (a), compute $y(n)$ for $n = 1, 2, 3, 4, 5$ when $y(0) = -1$ and $r(n) = 1$ for $n \geq 0$.

6.2   Repeat Problem 6.1 for the system given by

$$y(n+1) - 2y(n-1) + y(n-2) = c(n) - 4c(n-1)$$

6.3   Again consider the system given in Problem 6.1.

(a) Using the MPC approach, design an optimal controller so that the criterion (6.18) is minimized. Express your answer by giving the equation for the control $c(n)$.

(b) For your result in part (a), compute $y(n)$ for $n = 1, 2, 3, 4, 5$ when $h = 1$, $y(0) = -1$, and $r(n) = 1$ for $n \geq 0$.

6.4   For the disturbance estimator given by (6.14), verify that $\widehat{w}(n) = w(n)$ for all $n \geq 1$ when $w(n)$ is constant.

6.5 The system in Problem 6.1 is now subjected to a disturbance $w(n)$ so that the input–output equation becomes

$$y(n+1) + y(n) - y(n-1) = c(n-1) + w(n)$$

(a) By generalizing the approach given in this chapter, derive an expression for the estimate $\hat{w}(n)$ of $w(n)$.

(b) Using your result in part (a) and the MPC approach, design a controller so that $y(n)$ tracks a reference $r(n)$. Express your answer by giving the equation for the control $c(n)$.

(c) For your result in part (b), compute $y(n)$ for $n = 1, 2, 3, 4, 5$ when $w(n) = 3$ for all $n \geq 0$, $y(0) = -1$, and $r(n) = 1$ for all $n \geq 0$.

6.6 A two-input/two-output system is given by

$$y(n+1) = Ay(n) + Bc(n)$$

where

$$A = \begin{bmatrix} 0 & 1 \\ 1 & 0 \end{bmatrix} \text{ ανδ } B = \begin{bmatrix} -1 & 1 \\ 2 & 1 \end{bmatrix}$$

(a) Using the MPC approach, design a control so that $y(n)$ tracks a reference $r(n)$. Express your answer by giving the equation for the control $c(n)$.

(b) For your result in part (a), compute $y(n)$ for $n = 1, 2, 3$ when

$$y(0) = \begin{bmatrix} 1 \\ -1 \end{bmatrix} \text{ and } r(n) = \begin{bmatrix} 1 \\ 1 \end{bmatrix} \text{ for } n \geq 0$$

6.7 A system is given by $y(n+1) = ay(n) + bc(n)$, where a and b are unknown. When $y(0) = 1$, $c(n) = 0$ for $n < 0$, and $c(n) = 1$ for $n \geq 0$, it is known that $y(1) = 0.5$, $y(2) = 1.25$, $y(3) = 1.625$, $y(4) = 1.8125$, and $y(5) = 1.90625$.

(a) Using the estimation approach given in this chapter, compute the estimates $\hat{a}(n)$, $\hat{b}(n)$ of a, b for $n = 3, 4, 5$.

(b) Using your result in part (a) and the MPC approach, compute a control $c(n)$ for $n = 3, 4, 5$ that forces $y(n)$ to track the reference $r(n)$ = 1 for $n \geq 0$. Express your answer by giving the equation for the control values $c(3)$, $c(4)$, and $c(5)$.

6.8   For the system in Problem 6.7, use direct adaptive control to generate the control values $c(3)$, $c(4)$, and $c(5)$ for tracking the reference $r(n) = 1$ for $n \geq 0$. Express your answer by giving the equation for the control values $c(3)$, $c(4)$, and $c(5)$.

Chapter 7

Discrete Logic Control

In Chapters 4 through 6 we studied continuous-variable time-driven control where the inputs and outputs of the controllers are continuous variables that are functions of time. In this chapter we study discrete logic controllers where the inputs and outputs are discrete variables that change values as a result of the occurrence of events.

A multiple-input/multiple-output (MIMO) discrete logic controller is illustrated in Figure 7.1. Here I_1, I_2, ..., I_p are the inputs and Y_1, Y_2, ..., Y_m are the outputs. The I_i and Y_i are discrete-valued variables; that is, they have a finite number of possible values. As first noted in Chapter 2, the controller is called a discrete logic controller since the control action is given by logic (if–then) equations with discrete values. Note also that we are now using Y_i to denote the controller outputs, whereas before we were using Y or y to denote the output of the process being controlled.

In our study of discrete logic control, we will consider only the binary case where the controller inputs I_i and outputs Y_i take on only the two values 0 and 1, where 0 denotes "off" and 1 denotes "on." It turns out that any discrete logic controller can be described in terms of binary variables. Hence, there is no loss of generality in considering only the binary-variable case.

The controller inputs I_i represent the on/off status of input devices such as on/off switches, pushbutton switches, limit switches, and proximity switches. Examples of such devices are given in examples in this chapter and the next one. The controller outputs are on/off signals that are sent to "output devices" such as lights, motor starters, pumps, and solenoids that drive valves. The input and output devices are often referred to as *field devices* since they are typically lo-cated at a distance from the controller, so the input and output devices may be in

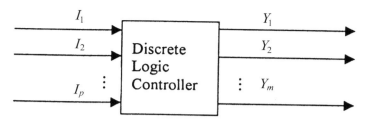

Figure 7.1   MIMO discrete logic controller.

the "field." The input and output devices may be components of some process or machine, or they may be components of a distributed control system (DCS).

The control problem of interest is the design and implementation of a discrete logic controller that carries out some desired control action or some sequence of desired control actions. The control action is achieved by controlling the on/off status of output devices using measurements of the on/off status of the input devices. We can approach the problem of controller design by using a state variable approach. We begin by considering state diagrams.

State Diagram Representation

A discrete logic controller can be modeled in terms of a state diagram. State diagrams are becoming very popular because they provide a convenient framework for representing the "dynamics" of discrete event processes or systems including discrete logic controllers.

A state diagram is a type of graph consisting of nodes corresponding to the states of the process under study, and with connections between the nodes corresponding to the conditions (viewed as events) that cause transitions from one state to another. The states are often defined in terms of the on/off status of output devices. Two examples are given next. Suppose that the objective is to design a discrete logic controller that keeps the level of fluid in a tank at a desired level L. The tank with input valve and a level limit switch is illustrated in Figure 7.2. The input valve controls the flow of fluid into the tank. The status of the valve is denoted by V where $V = 0$ when the valve is closed (off) and $V = 1$ when the valve is open (on). The level limit switch determines whether the level in the tank is below or above the desired level L. The status of the switch is denoted by LLS, where $LLS = 0$ when the level is below L, and $LLS = 1$ when the level is at or above L.

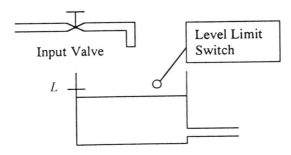

Figure 7.2 Tank with input valve and level limit switch.

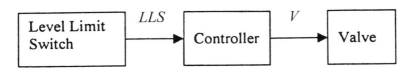

Figure 7.3 Controller with connections to the input and output devices.

In this example, the input device is the level limit switch and the output device is the input valve. The connection of these devices to the controller is shown in Figure 7.3. The desired control action is to have the controller send the command $V = 1$ to open the valve or keep the valve open when the level of fluid in the tank is less than L ($LSS = 0$), and to have the controller send the command $V = 0$ to close the valve or keep the valve closed when the level is equal to or greater than L ($LLS = 1$).

To describe this control action using a state diagram, we first need to define the states. The obvious choice for the states are "valve closed" ($V = 0$) and "valve open" ($V = 1$). When the valve is closed, the input value $LLS = 0$ will cause the valve to open, and thus $LLS = 0$ causes a state transition from "valve closed" to "valve open." When the valve is open, the input value $LLS = 1$ will cause the valve to close, and thus $LLS = 1$ causes a state transition from "valve open" to "valve closed." The state diagram shown in Figure 7.4 describes this control action.

It is important to note that in Figure 7.4 we are showing only those transitions from one state to another state. There are also transitions from each state back to that state. In particular, if the valve is closed and $LLS = 1$, the valve remains closed, and if the valve is open and $LLS = 0$, the valve remains open. In the design of the controller, it turns out that it is not necessary to consider the transitions from a state back to itself, and thus any such transitions will not be shown in the state diagram.

The states "valve closed" and "valve open" shown in Figure 7.4 can be represented in terms of a state variable X, where $X = 0$ when the valve is closed and $X = 1$ when the valve is open. Expressing the state diagram in terms of X yields the result shown in Figure 7.5. Note that by definition of X, we have that $X = V$, and thus the state variable X is equal to the status of the output device (i.e., the valve).

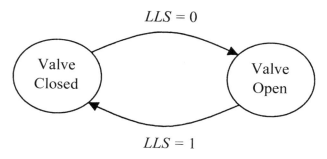

Figure 7.4   State diagram describing desired control action.

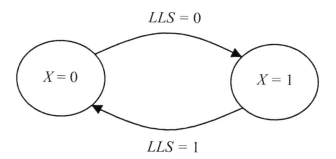

Figure 7.5   State diagram given in terms of the state variable X.

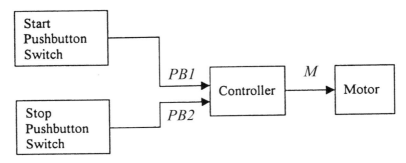

Figure 7.6 Connection of pushbutton switches and motor to controller.

For a second example, suppose that the goal is to control the starting and stopping of a motor using two pushbutton switches. The status of the motor is denoted by M with $M = 0$ when the motor is off and $M = 1$ when the motor is on. The status of the pushbutton switches is denoted by $PB1$ and $PB2$, where 0 means off and 1 means on. It is assumed that each switch has momentary closure; that is, the switch closes for a short period of time after the button is pressed and then opens.

Note that in this example the pushbutton switches are the input devices and the motor is the output device. The connection of the switches and motor to the controller is shown in Figure 7.6. In terms of this configuration, the desired control action is as follows:

1. The controller sends the command $M = 1$ to turn the motor on if it is off and the start pushbutton switch is pressed ($PB1 = 1$), or to keep the motor on if it is already on and the stop pushbutton switch has not been pressed ($PB2 = 0$).

2. The controller sends the command $M = 0$ to turn the motor off if it is on and the stop pushbutton switch is pressed ($PB2 = 1$), or to keep the motor off if it is already off and the start pushbutton switch is not pressed ($PB1 = 0$).

Here we are assuming that the two pushbutton switches are not pressed at the same time. Later we will see how to handle this case.

The states for this control action are "motor off" and "motor on," and in terms of these states the state diagram is shown in Figure 7.7. Note again that we are not showing the transitions from a state back to itself. If we define the state variable X, where $X = 0$ when the motor is off and $X = 1$ when the motor is

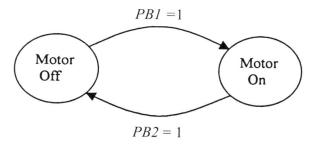

Figure 7.7   State diagram for motor control.

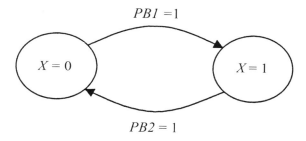

Figure 7.8   State diagram in terms of X.

on, we have that $X = M$, and the state diagram expressed in terms of X is given in Figure 7.8.

Example Having Two State Variables

In each of the preceding two examples, there was only one state variable. We can generate an example with two state variables by making the level-control problem more complicated (see Figure 7.2): Suppose that a pump must be turned on to force the liquid through the input valve. We let P denote the status of the pump, where 0 means off and 1 means on. We also have the constraint that the pump can be on only when the input valve is open, otherwise, the pump may burn out.

The desired control action is as follows

1.   Open the valve if it is closed and the level of liquid in the tank is less than the desired level L ($LLS = 0$), or keep the valve open if $LLS = 0$.

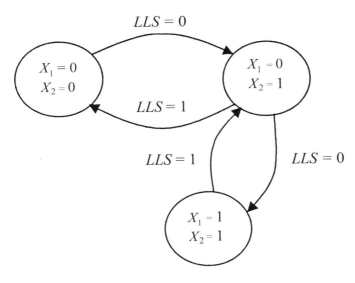

Figure 7.9 State diagram in terms of the state variables $X_1 = P$ and $X_2 = V$.

2. Close the valve if it is open and the level of liquid in the tank is equal to or greater than the desired level L ($LLS = 1$), or keep the valve closed if $LLS = 1$.

3. Turn the pump on if it is off and the valve is open and $LLS = 0$, or keep the pump on if it is already on and the valve is open and $LSS = 0$.

3. Turn the pump off if it is on and $LLS = 1$, or keep the pump off if $LLS = 1$.

This control action can be described in terms of two state variables X_1 and X_2, with $X_1 = P$ and $X_2 = V$. In terms of X_1 and X_2 the states are

$$X_1 = 0, \quad X_2 = 0 \ (\text{pump off, valve closed}),$$
$$X_1 = 0, \quad X_2 = 1 \ (\text{pump off, valve open}),$$
$$X_1 = 1, \quad X_2 = 1 \ (\text{pump on, valve open}).$$

Note that the state $X_1 = 1$, $X_2 = 0$ (pump on, valve closed) is not allowed since the pump must be off when the input valve is closed. Then in terms of the state variables X_1 and X_2, the state diagram describing the desired control action is shown in Figure 7.9.

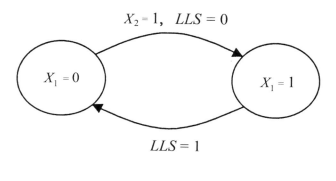

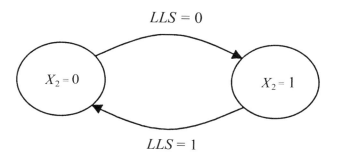

Figure 7.10   State diagrams for each of the state variables, X_1 and X_2

Instead of using the state diagram in Figure 7.9, we can express the desired control action in terms of two state diagrams, one for each of the state variables X_1 and X_2 : Starting with X_1, we determine the conditions that cause X_1 to transition from $X_1 = 0$ to $X_1 = 1$. Since $X_1 = P$, it is clear from the control statements that X_1 goes from $X_1 = 0$ (pump off) to $X_1 = 1$ (pump on) when $X_2 = 1$ (input valve is open) and $LLS = 0$. We indicate these conditions by putting $X_2 = 1$ and $LLS = 0$ along the arc from $X_1 = 0$ to $X_1 = 1$ in the top portion of the diagram in Figure 7.10. We then determine the conditions that cause X_1 to transition from $X_1 = 1$ to $X_1 = 0$. Again from the control statements, this transition occurs when $LLS = 1$, which is indicated on the arc from $X_1 = 1$ to $X_1 = 0$ in the top portion of Figure 7.10.

Via a similar procedure, we can determine the conditions that cause the state variable X_2 to transition from $X_2 = 0$ to $X_2 = 1$, and from $X_2 = 1$ to $X_2 = 0$. This results in the state diagram shown in the bottom part of Figure 7.10. Note that the transitions for the state variable X_2 do not depend on the state variable

X_1; whereas, the transition from $X_1 = 0$ to $X_1 = 1$ does depend on X_2. Hence, in this example X_2 is independent of X_1, but X_1 depends on X_2.

It turns out that the diagram in Figure 7.10 provides a much better framework than the diagram in Figure 7.9 for determining Boolean logic equations for the control action. The process of generating the logic equations is pursued in the following section.

Boolean Logic Equations

The design of a discrete logic controller can be carried out by generating Boolean logic equations that describe the desired control action. Boolean logic equations can be determined in a systematic manner from the statements comprising the desired control action by first defining state variables and determining the state transitions for each of the state variables. To describe the procedure for generating Boolean logic equations, we first need to review the elements of Boolean algebra and state variable representations.

Boolean Algebra

Let A and B be binary variables; that is, $A, B = 0$ or 1. We view A and B as Boolean variables where $A = 1$ ($B = 1$) means that A is true (resp., B is true), and $A = 0$ ($B = 0$) means that A is not true (resp., B is not true). The addition A+B means that either A or B is true and is given by

$$A + B = 0 \quad \text{when} \quad A = 0 \quad \text{and} \quad B = 0$$
$$A + B = 1 \quad \text{otherwise}$$

The multiplication AB means that both A and B are true and is given by

$$AB = 1 \quad \text{when} \quad A = 1 \quad \text{and} \quad B = 1$$
$$AB = 0 \quad \text{otherwise}$$

There is also the *not operation*, denoted by $\overline{A}$, which stands for "not A," and which is defined by

$$\overline{A} = 1 \quad \text{when} \quad A = 0$$
$$\overline{A} = 0 \quad \text{when} \quad A = 1$$

The addition and multiplication operations have a number of properties that are very useful in working with Boolean equations. We list some of these properties below:

$$A + 0 = A, \quad A + 1 = 1, \quad A + A = A, \quad A + \overline{A} = 1,$$
$$A(0) = 0, \quad A(1) = A, \quad AA = A, \quad A\overline{A} = 0, \quad A(B + A) = A,$$
$$\overline{A + B} = \overline{A}\ \overline{B}, \quad \overline{AB} = \overline{A} + \overline{B}$$

We will need to consider functions of two or more Boolean variables; for example, suppose that $f(A,B)$ is a function of the Boolean variables A and B. Any such function $f(A,B)$ can be expressed in terms of a sum of products of A, $\overline{A}$, B, $\overline{B}$ taken two at a time; that is,

$$f(A,B) = aAB + bA\overline{B} + c\overline{A}B + d\overline{A}\,\overline{B}$$

where a, b, c, d are equal to 0 or 1. Note that

$$a = f(A,B) \quad \text{with } A = 1, B = 1$$
$$b = f(A,B) \quad \text{with } A = 1, B = 0$$
$$c = f(A,B) \quad \text{with } A = 0, B = 1$$
$$d = f(A,B) \quad \text{with } A = 0, B = 0$$

For example, suppose

$$f(A,B) = AB(A+B) + B(AB) + A$$

Then

$$a = (1)(1)(1+1)+(1)(1)(1)+1 = 1$$
$$b = (1)(0)(1+0)+(0)(1)(0)+1 = 1$$
$$c = (0)(1)(0+1)+(1)(0)(1)+0 = 0$$
$$d = (0)(0)(0+0)+(0)(0)(0)+0 = 0$$

and thus

$$f(A,B) = AB + A\overline{B} = A$$

It also follows from the sum of products form that any function $f(A_1, A_2, \ldots, A_q)$ of q Boolean variables $A_1, A_2, \ldots, A_q$ can be expressed in the form

$$f(A_1, A_2, \ldots, A_q) = \alpha(A_1, A_2, \ldots, A_{q-1})A_q + \beta(A_1, A_2, \ldots, A_{q-1})\overline{A_q} \quad (7.1)$$

where $\alpha(A_1, A_2, \ldots, A_{q-1})$ and $\beta(A_1, A_2, \ldots, A_{q-1})$ are functions of $A_1, A_2, \ldots, A_{q-1}$. Note that if

$$\alpha(A_1, A_2, \ldots, A_{q-1}) = \beta(A_1, A_2, \ldots, A_{q-1}) \quad (7.2)$$

from (7.1) we see that $f(A_1, A_2, \ldots, A_q) = \alpha(A_1, A_2, \ldots, A_{q-1})$, and thus $f(A_1, A_2, \ldots, A_q)$ is independent of A_q, that is, the value of $f(A_1, A_2, \ldots, A_q)$ is independent of the value of A_q. The condition (7.2) is necessary and sufficient for $f(A_1, A_2, \ldots, A_q)$ to be independent of A_q.

State Variable Representation

A discrete logic controller can be represented in terms of a state variable representation given in terms of binary-valued state variables denoted by $X_1, X_2, \ldots, X_N$. For a controller with inputs $I_1, I_2, \ldots, I_p$ and outputs $Y_1, Y_2, \ldots, Y_m$, the dynamics of the controller are given by the state equations

$$X_i^+ = f_i(X_1, X_2, \ldots, X_N, I_1, I_2, \ldots, I_p), \quad i = 1, 2, \ldots, N \quad (7.3)$$

$$Y_i = h_i(X_1, X_2, \ldots, X_N), \quad i = 1, 2, \ldots, m \quad (7.4)$$

where X_i^+ specifies the next value of the state variable X_i; the f_i are functions of the variables $X_1, X_2, \ldots, X_N, I_1, I_2, \ldots, I_p$; and the h_i are functions of $X_1, X_2, \ldots, X_N$. Here all the variables are viewed as Boolean variables, and thus (7.3) and (7.4) are Boolean logic equations.

It is important to emphasize that a change in state in the controller defined by (7.3) and (7.4) is a result of the event that one or more of the inputs $I_1, I_2, \ldots, I_p$ change value, so state transitions are event driven, not time driven. Hence, the controller defined by (7.3) and (7.4) is event driven, not time driven.

In the case when $p=m=N=1$, so that there is a single input I, single output Y, and single state variable X, the state variable representation is

$$X^+ = f(X, I) \tag{7.5}$$

$$Y = h(X) \tag{7.6}$$

This controller has two states given by $X = 0$ and $X = 1$. The next state X^+ can be expressed in the form

$$X^+ = \alpha(I)X + \beta(I)\bar{X} \tag{7.7}$$

where $\alpha(I)$ and $\beta(I)$ are functions of the input I given by

$$\alpha(I) = f(1, I) = f(X, I) \quad \text{with} \ X = 1$$

$$\beta(I) = f(0, I) = f(X, I) \quad \text{with} \ X = 0$$

Note that if $\alpha(I) = \beta(I)$, then from (7.7) we have that

$$X^+ = \alpha(I)\left[X + \bar{X}\right] = \alpha(I)$$

In this case we see that the next state X^+ depends only on the input I, and not on the previous state X. Such a controller is said to be *memoryless*, since the value of the next state does not depend on the value of the previous state. In the general case given by (7.3) and (7.4), the controller is memoryless if and only if the next state X_i^+ is a function of only the inputs $I_1, I_2, \ldots, I_p$; that is, X_i^+ is independent of the values of the previous states (the X_i).

From (7.7), we see that if $X = 0$, the next state is $X^+ = \beta(I)$, and if $X = 1$, the next state is $X^+ = \alpha(I)$. Since $X^+ = \beta(I)$ when $X = 0$, the condition $\beta(I) = 1$ sets the value of X to 1 starting from $X = 0$. As a result, $\beta(I)$ can be viewed as a "set operation," which we can express in the form

$$\text{Set}_X = \beta(I) \tag{7.8}$$

Since $X^+ = \alpha(I)$ when $X = 1$, the condition $\alpha(I) = 0$, or $\overline{\alpha(I)} = 1$, resets the value of X to 0 starting from $X = 1$. Hence, $\alpha(I)$ can be viewed as a "reset operation," which we can express in the form

$$\text{Reset}_X = \overline{\alpha(I)} \text{ or } \overline{\text{Reset}_X} = \alpha(I) \tag{7.9}$$

In terms of the set and reset operations, the next state X^+ given by (7.7) can be written in the form

$$X^+ = (\overline{\text{Reset}_X})X + (\text{Set}_X)\overline{X} \tag{7.10}$$

If Set_X and Reset_X cannot both have the value 1 at the same time, (7.10) is equivalent to the standard *reset/set flip flop* equation given by

$$X^+ = (\overline{\text{Reset}_X})\left[X + \text{Set}_X\right] \tag{7.11}$$

The only difference between (7.10) and (7.11) is the value of X^+ when $\text{Set}_X = \text{Reset}_X = 1$ and $X = 0$: Evaluating (7.10) gives $X^+ = 1$ and evaluating (7.11) gives $X^+ = 0$. If $X^+ = 0$ is an "acceptable" value of X^+ when $\text{Set}_X = \text{Reset}_X = 1$ and $X = 0$, then (7.11) may be taken as the state equation of the controller.

Two State Variables

Now suppose that the controller under study has two state variables X_1 and X_2, a single input I, and a single output Y. In this case, the state variable model is

$$X_i^+ = f_i(X_1, X_2, I), \quad i = 1, 2$$

$$Y = h(X_1, X_2)$$

This controller has four states that can be denoted by the two-element strings:

$$X_1X_2 = 00$$

$$X_1X_2 = 01$$

$$X_1X_2 = 11$$

$$X_1X_2 = 10$$

Using the sums of products form, the next states X_1^+ and X_2^+ can be written in the form

$$X_1^+ = \alpha_1(X_2, I)X_1 + \beta_1(X_2, I)\overline{X_1} \tag{7.12}$$

$$X_2^+ = \alpha_2(X_1, I)X_2 + \beta_2(X_1, I)\overline{X_2} \tag{7.13}$$

where $\alpha_1(X_2,I)$, $\beta_1(X_2,I)$, $\alpha_2(X_1,I)$, $\beta_2(X_1,I)$ are functions given by

$$\alpha_1(X_2, I) = f_1(1, X_2, I)$$

$$\beta_1(X_2, I) = f_1(0, X_2, I)$$

$$\alpha_2(X_1, I) = f_2(X_1, 1, I)$$

$$\beta_2(X_1, I) = f_2(X_1, 0, I)$$

We can express the state transitions for the state variables X_1 and X_2 in terms of set and reset operations, as we did earlier for the single state variable case. First, note that when $X_1 = 0$, from (7.12) we have that

$$X_1^+ = \beta_1(X_2, I)$$

Hence, the condition $\beta_1(X_2,I) = 1$ sets X_1 to 1 from 0, which we can express in the form

$$\text{Set}_{X_1} = \beta_1(X_2, I) \tag{7.14}$$

Equation (7.14) is the operation for setting X_1 to 1 from 0.

Since $X_1^+ = \alpha_1(X_2,I)$ when $X_1 = 1$, the condition $\alpha_1(X_2,I) = 1$ or $\alpha_1(X_2,I) = 0$ resets the value of X_1 to 0 starting from $X_1 = 1$. Thus, $\alpha_1(X_2,I)$ can be viewed as a reset operation, which we can express in the form

$$\text{Reset}_{X_1} = \overline{\alpha_1(X_2, I)} \quad \text{or} \quad \text{Reset}_{X_1} = \alpha_1(X_2, I) \tag{7.15}$$

Then using (7.14) and (7.15) in (7.12), we have that the next state X_1^+ can be expressed in the form

$$X_1^+ = (\overline{\text{Reset}_{X_1}})X_1 + (\text{Set}_{X_1})\overline{X_1} \qquad (7.16)$$

Via a similar analysis for the state variable X_2, we have the set and reset operations:

$$\text{Set}_{X_2} = \beta_2(X_1, I) \qquad (7.17)$$

$$\text{Reset}_{X_2} = \overline{\alpha_2(X_1, I)} \quad \text{or} \quad \overline{\text{Reset}_{X_2}} = \alpha_2(X_1, I) \qquad (7.18)$$

Using (7.17) and (7.18) in (7.13), we have that the next state X_2^+ can be expressed in the form

$$X_2^+ = (\overline{\text{Reset}_{X_2}})X_2 + (\text{Set}_{X_2})\overline{X_2} \qquad (7.19)$$

To summarize, we have shown that a controller with two state variables given by the state equations (7.12) and (7.13) can be described in terms of set and reset operations, which results in the state equations given by (7.16) and (7.19). If Set_{X_i} and Reset_{X_i} cannot both be 1 at the same time or if X_i^+ can be set equal to 0 when $\text{Set}_{X_i} = \text{Reset}_{X_i} = 1$ and $X_i = 0$, the state equations can be expressed in the reset/set flip flop equation form:

$$X_1^+ = (\overline{\text{Reset}_{X_1}})\left[X_1 + \text{Set}_{X_1}\right] \qquad (7.20)$$

$$X_2^+ = (\overline{\text{Reset}_{X_2}})\left[X_2 + \text{Set}_{X_2}\right] \qquad (7.21)$$

N State Variables

The above results for the case of two state variables generalize to discrete logic controllers with N state variables X_1, X_2, ..., X_N. Given any such controller, for each fixed value of $i = 1, 2, ..., N$, we define Set_{X_i} to be the sum of all those conditions whereby X_i is set to 1 from 0, and we define Reset_{X_i} to be the sum of all those conditions whereby X_i is reset to 0 from 1. Then the next state X_i^+ is given by

$$X_i^+ = (\overline{\text{Reset}_{X_i}})X_i + (\text{Set}_{X_i})\overline{X_i} \qquad (7.22)$$

If Set_{X_i} and Reset_{X_i} cannot both be 1 at the same time or if X_i^+ can be set equal to 0 when $\text{Set}_{X_i} = \text{Reset}_{X_i} = 1$ and $X_i = 0$, the state equation for X_i^+ is given by

$$X_i^+ = (\overline{\text{Reset}_{X_i}})\left[X_i + \text{Set}_{X_i}\right] \qquad (7.23)$$

Carrying out this construction for $i = 1, 2, \ldots, N$ results in the state equations for the system.

Generation Of Boolean Logic Equations

Given the statements for a desired control action, the first step in constructing the Boolean logic state equations for the discrete logic controller is the definition of the state variables $X_1, X_2, \ldots, X_N$. If the outputs of the controller are $Y_1, Y_2, \ldots, Y_m$, then we can take X_i to be equal to Y_i for $i = 1, 2, \ldots, m$. In some cases, this definition of the state variables will work without having to define additional state variables. However, in general the number N of state variables that are required is greater than the number m of output variables. In particular, as will be seen in the next chapter, the construction of the state equations may require "internal coils," which will result in a larger value of N than m.

We begin by considering a discrete logic controller with a single state variable $X = Y$, where Y is the output variable. In this case there are two states given by $X = 0$ and $X = 1$, and the general form of the state diagram is shown in Figure 7.11. The variables A and B in Figure 7.11 are functions of the inputs to the controller.

The set and reset conditions for X can be determined directly from the state diagram in Figure 7.11. In this case, X is set to 1 from 0 when $A = 1$, and X is reset to 0 from 1 when $B = 1$. Hence,

$$\text{Set}_X = A \quad \text{and} \quad \text{Reset}_X = B$$

and using (7.10) results in the state equation

$$X^+ = \overline{B}X + A\overline{X} \tag{7.24}$$

As noted before, if both A and B cannot be 1 at the same time or if X^+ can be set equal to 0 when $A = B = 1$ and $X = 0$, the state equation becomes

$$X^+ = \overline{B}(X + A) \tag{7.25}$$

The state diagram corresponding to the state equation (7.25) is given in Figure 7.12. Note the difference between the state diagrams in Figures 7.11 and 7.12: In (7.12) the transition from $X = 0$ to $X = 1$ occurs if $A = 1$ and $B = 0$; whereas, in Figure 7.11 this transition occurs if $A = 1$ (so that the value of B does not matter).

As an application of this result, again consider the motor with the start and stop pushbutton switches (see Figure 7.6). Recall that $PB1$ and $PB2$ are the on/off status of the switches and the state variable X was defined to be equal to M, where M is the on/off status of the motor. Then comparing the state diagram in Figure 7.8 with the diagram in Figure 7.11 reveals that $A = PB1$ and $B = PB2$. Hence, using (7.25) we have the state equation

$$X^+ = (\overline{PB2})(X + PB1) \tag{7.26}$$

From (7.26), we see that $X^+ = 0$ if $PB1 = PB2 = 1$ and $X = 0$. Thus, if the motor is off and both pushbuttons are pressed at the same time, the motor remains off.

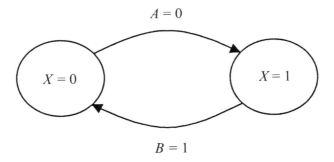

Figure 7.11 State diagram in single state variable case.

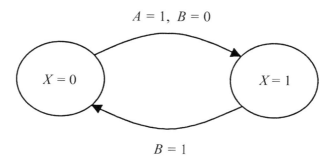

Figure 7.12   Modified state diagram in single state variable case.

     We should also point out that the discrete logic controller given by (7.26) has memory since the next value of the state depends in general on the previous value of the state. In this example, the memory is needed to "latch" the motor so that it stays on after the momentary closure of the start pushbutton switch. More precisely, if the motor is on ($X = 1$), the motor stays on ($X^+ = 1$) if the stop pushbutton switch is not pressed ($PB2 = 0$), even though $PB1 = 0$.

Example with Two State Variables

Consider the liquid level control problem with the state diagram shown in Figure 7.9. Recall that the status of the level limit switch, pump, and input valve are denoted by LLS, P, and V, respectively. Then with the state variables defined by $X_1 = P$ and $X_2 = V$, from the state diagrams for X_1 and X_2 given in Figure 7.10, we have that the set and reset conditions are

$$\text{Set}_{X_1} = (\overline{LLS})X_2, \quad \text{Reset}_{X_1} = LLS$$

$$\text{Set}_{X_2} = \overline{LLS}, \quad \text{Reset}_{X_2} = LLS$$

and thus using (7.22), we have that the state equations are

$$X_1^+ = (\overline{LLS})X_1 + (\overline{LLS})X_2\,\overline{X_1} \tag{7.27}$$

$$X_2^+ = (\overline{LLS})X_2 + (\overline{LLS})\,\overline{X_2} \tag{7.28}$$

In this case, Set_{X_i} and Reset_{X_i} cannot both be 1 at the same time, and thus (7.27) and (7.28) are equivalent to the reset/set flip flop equations

$$X_1^+ = (\overline{\mathrm{Reset}_{X_1}})\left[X_1 + \mathrm{Set}_{X_1}\right] = (\overline{LLS})\,[X_1 + (\overline{LLS})X_2] \quad (7.29)$$

$$X_2^+ = (\overline{\mathrm{Reset}_{X_2}})\left[X_2 + \mathrm{Set}_{X_2}\right] = (\overline{LLS})\,[X_2 + (\overline{LLS})] \quad (7.30)$$

Equations (7.29) and (7.30) simplify to

$$X_1^+ = (\overline{LLS})(X_1 + X_2) \qquad (7.31)$$

$$X_2^+ = \overline{LLS} \qquad (7.32)$$

Note that by (7.31), the pump is on if the level of the liquid is less than the desired level ($LLS = 0$) and the input valve is on or the pump is already on. By (7.32), the input valve is on if $LLS = 0$.

A Second Two State Variable Example

We conclude this chapter with another example having two state variables. A much more complicated example with five state variables is considered in the next chapter.

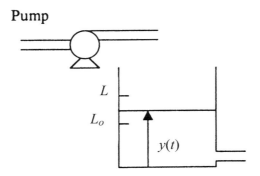

Figure 7.13 Tank with two-speed pump.

Consider the tank with pump shown in Figure 7.13. The pump has an on/off switch whose status is denoted by S where $S = 0$ when the pump off and $S = 1$ when the pump is on. When the pump is on it can operate at either slow speed or fast speed. When the pump is turned on, the level $y(t)$ of fluid in the tank could have any value between 0 and L, where L is the desired level. The objective is to design a discrete logic controller to operate the pump so that it's on fast when $y(t) < L_o$ and on slow when $L_o \leq y(t) < L$, where L_o is an intermediate level, and the pump is off when $y(t) \geq L$ or when $S = 0$. There are two level limit switches with the status of the first switch denoted by $LLS1$ and the status of the second switch denoted by $LLS2$. The level limit switches provide the following information:

$$LLS1 = 0 \text{ when } y(t) < L_o \quad \text{and} \quad LLS1 = 1 \text{ when } y(t) \geq L_o$$

$$LLS2 = 0 \text{ when } y(t) < L \quad \text{and} \quad LLS2 = 1 \text{ when } y(t) \geq L$$

The desired control action is as follows:

1. Turn the pump on fast if $S = 1$ and $LLS1 = 0$.
2. Turn the pump on slow if $S = 1$, $LLS1 = 1$, and $LLS2 = 0$.
3. Turn the pump off if $S = 0$ or $LLS2 = 1$.

The controller has three inputs equal to S, $LLS1$, and $LLS2$, and it has two outputs Y_1 and Y_2 defined by

$Y_1 = 1$ when the pump is on fast, $Y_1 = 0$ when the pump is not on fast

$Y_2 = 1$ when the pump is on slow, $Y_2 = 0$ when the pump is not on slow

We can define the state variables to be equal to the outputs; that is, we take $X_1 = Y_1$ and $X_2 = Y_2$. Then since Y_1 and Y_2 are independent; that is, the value of Y_1 does not depend on the value of Y_2, and vice versa, X_1 and X_2 are independent. Hence, the state diagram for X_1 (resp., X_2) does not depend on X_2 (resp., X_1). The state diagrams for X_1 and X_2 are shown in Figure 7.14.

From Figure 7.14, we have that the set and reset operations for each state variable are

$$\text{Set}_{X_1} = S(\overline{LLS1}) \quad \text{and} \quad \text{Reset}_{X_1} = \overline{S} + LLS2 + S(LLS1)(\overline{LLS2})$$

$$\text{Set}_{X_2} = S(LLS1)(\overline{LLS2}) \quad \text{and} \quad \text{Reset}_{X_2} = \overline{S} + LLS2 + S(\overline{LLS1})$$

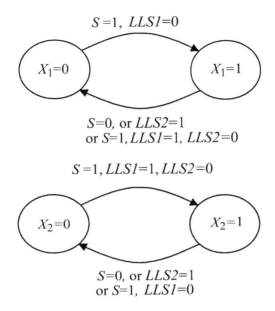

Figure 7.14 State diagram for tank with two-speed pump.

Then using (7.23) results in the state equations

$$X_1^+ = S(\overline{LLS2})(\overline{S} + \overline{LLS1} + LLS2)[X_1 + S(\overline{LLS1})]$$

$$= S(\overline{LLS2})(\overline{LLS1})[X_1 + S(\overline{LLS1})]$$

$$= S(\overline{LLS2})(\overline{LLS1}) \tag{7.33}$$

and

$$X_2^+ = S(\overline{LLS2})(\overline{S} + LLS1)[X_2 + S(LLS1)(\overline{LLS2})]$$

$$= S(\overline{LLS2})(LLS1)[X_2 + S(LLS1)(\overline{LLS2})]$$

$$= S(\overline{LLS2})(LLS1) \tag{7.34}$$

From (7.33) and (7.34), we see that both X_1^+ and X_2^+ are independent of X_1 and X_2, and thus the controller is memoryless.

Problems

7.1.   A discrete logic controller controls a machine so that it starts operation when a start pushbutton switch is pressed and there is no fault. The status of the machine and start pushbutton switch are denoted by M and *PB1*, respectively, and the status of the fault is denoted by F, where $F = 0$ means no fault and $F = 1$ means there is a fault. The controller turns off the machine when the stop pushbutton switch is pressed or if there is a fault. The status of the stop pushbutton switch is denoted by *PB2*. It is assumed that a fault is detected by a sensor.

   a.   Define the inputs, outputs, and states of the controller.

   b.   Generate a single state diagram for the controller.

   c.   Generate a state diagram for each state variable of the controller.

   d.   Using your result in part (c), generate Boolean logic equations for the control action. Express the equations in the simplest possible form

7.2.   A discrete logic controller for an oven operates as follows: If the oven on switch is activated, the oven door is closed, and the temperature is below a desired value, the controller turns the heater on. If the heater is on, or when the temperature is above the desired value and the door is closed, the controller turns the fan on. If the oven light switch is on or the door is open, the controller turns the oven light on. The oven has an oven door limit switch whose status is denoted by DLS, where $DLS = 0$ means the door is open and $DLS = 1$ means the door is closed. The oven also has a temperature limit switch whose status is denoted by TLS, where $TLS = 0$ when the temperature is blow the desired value and $TLS = 1$ when the temperature is at or above the desired value. The status of the heater, fan, light, oven on switch, and light switch is denoted by H, F, L, S, and LS, respectively.

   a.   Define the inputs, outputs, and states of the controller.

   b.   Generate a single state diagram for the controller.

c. Generate a state diagram for each state variable of the controller.

d. Using your result in part (c), generate Boolean logic equations for the control action. Express the equations in the simplest possible form

7.3. Liquid flows into a mixing tank through an input pipe with input pump, and liquid flows out of the tank through an output pipe with output pump. Each pump is either on or off. The tank also has other input and output pipes through which liquid is added or removed. The objective is to design a discrete logic controller so that the input and output pumps are turned on and off to maintain the level of the liquid in the tank between 6 and 8 ft. A level limit switch is closed ($LLS1 = 1$) when the liquid level is greater than or equal to 6 ft, and a second level limit switch is open ($LLS2 = 0$) when the liquid level is less than or equal to 8 ft. The initial level of the liquid in the tank may be less than 6 ft.

a. Define the inputs, outputs, and states of a discrete logic controller for the mixing tank.

b. Generate a single state diagram for the discrete logic controller.

c. Generate a state diagram for each state variable of the controller.

d. Using your result in part (c), generate Boolean logic equations for the control action. Express the equations in the simplest possible form

7.4. A discrete logic controller is given by the state diagram shown in Figure P7.4.

a. Generate a state diagram for each state variable of the controller.

b. Using your result in part (a), generate Boolean logic equations for the control action. Express the equations in the simplest possible form.

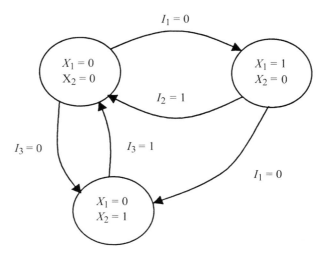

Figure P7.4

7.5   A discrete logic controller with inputs I_1 and I_2 is given by the state diagram shown in Figure P7.5.

   a.   Determine the Boolean logic equations that describe the controller. Express the equations in the simplest possible form.

   b.   Redraw the state diagram in terms of the states 000, 001, 010, 100, etc. Use only those states that are possible for the given control function.

7.6   The objective is to design a discrete logic controller for the processing of parts through two machines that are connected in series as shown in Figure P7.6. Each machine has an input position where the part is worked on, and an output position where the part is stored before it leaves the machine after processing. Limit switches are available for detecting the presence of a part at the input and output positions of each machine. A robot whose status is denoted by *R1* moves parts one at a time from a supply bin to the input position of the first machine. A robot whose status is denoted by *R2* moves parts one at a time from the output position of the first machine to the input position of the second machine. The status of the machines is denoted by *M1* and *M2*. When either robot is turned on, it carries out a preprogrammed motion corresponding to the desired trajectory and then turns off

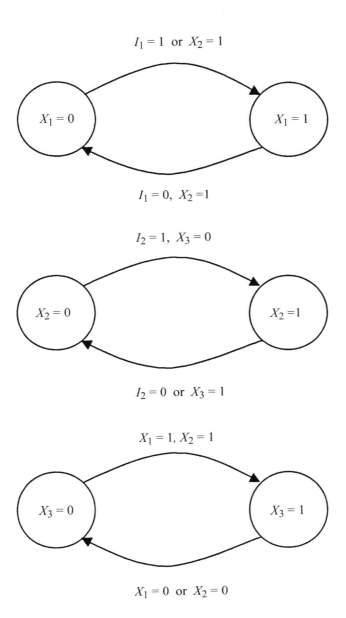

Figure P7.5

after the part it is moving has been placed in the desired position. When either machine is turned on, it carries out a prescribed operation which includes moving the part to the output position when it is done, and then turns off. The first robot moves a part from the supply bin to the input position of the first machine whenever both the input position and output positions of the first machine are empty. The first machine starts whenever there is a part in its input position with no part in its output position. The same is true for the second machine. The second robot moves a part from the output position of the first machine to the input position of the second machine whenever there is a part in the output position of the first machine and there are no parts in the input and output positions of the second machine. The values of *R1* and *R2* are equal to 1 when the robots are operating, otherwise they are 0, and *M1* and *M2* are equal to 1 when the machines are operating, otherwise they are zero.

a.   Define the inputs, outputs, and states of the controller,

b.   Generate a state diagram for each state variable of the controller.

c.   Using your result in part (b), generate Boolean logic equations for the control action. Express the equations in the simplest possible form.

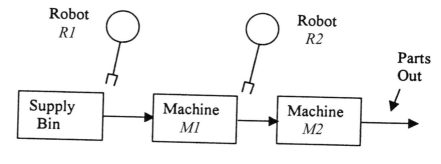

Figure P7.6

7.7   The objective is to design a discrete logic controller for the drilling of a part as follows. The part is first mounted into a fixture and then a start pushbutton is pressed to initiate the drilling sequence. The drill spindle turns on and the drill bit is moved down onto the part. After the desired depth is reached, the drill bit is retracted. The fixture is then moved to another drilling position and the drilling operation is repeated. Then the drill spindle turns off and the fixture moves back to the initial position. The components with status variables for the drilling operation are as follows:

- Start pushbutton (PB)
- Drill spindle motor ($M1$)
- Drill vertical movement motor ($M2$)
- Drill depth limit switch ($LS1$)
- Drill vertical position limit switch ($LS2$)
- Fixture movement motor ($M3$)
- First position fixture limit switch ($LS3$)
- Second position fixture limit switch ($LS4$)

a. Define the inputs, outputs, and states of the controller.

b. Generate a state diagram for each state variable of the controller.

c. Using your result in part (b), generate Boolean logic equations for the control action. Express the equations in the simplest possible form.

Chapter 8

Ladder Logic Diagrams and PLC Implementations

In the previous chapter the design of a discrete logic controller was approached by first determining the state transitions for the state variables, and then generating Boolean logic equations from the state transitions using set/reset operations. The resulting design can be implemented by using a PC with appropriate input/output (I/O) interfaces and with the Boolean logic equations programmed on the PC. However, it is common practice in industrial environments to implement a discrete logic controller on a programmable logic controller (PLC). A PLC implementation is often preferred over a PC implementation since PLCs have I/O interfaces built into the hardware and PLCs are in general better suited for the harsh conditions that may exist in an industrial setting.

PLCs came into existence in order to provide flexibility for modifying or expanding control functionality without having to make hardware modifications. Prior to PLCs, control was achieved using hardwired implementations that are not easily modified in order to accommodate changes in the control process. In contrast, the I/O connections to a PLC are easily altered or expanded and the control action is easily modified by simply reprogramming the PLC.

Hardwired control implementations are often specified in terms of an "electrical ladder logic diagram." As a result, when PLCs came into existence they were often programmed using a ladder logic diagram so that the programming would be familiar to anyone who had worked with electrical ladder logic diagrams. Although hardware implementations based on electrical ladder logic diagrams have for the most part been replaced by PLC implementations, the programming of a PLC is still often carried out by using a ladder logic diagram. The setting up and implementation of ladder logic diagrams are pursued in this chapter. Other approaches to programming PLCs, such as sequential function charts, are not considered (see Appendix A for an appropriate reference). We begin with electrical ladder logic diagrams.

Electrical Ladder Logic Diagrams

Suppose we want to control the on/off status of a light by using a switch. The status of the light is denoted by L and the status of the switch is denoted by S, where the light is on (off) when $L = 1$ ($L = 0$), and the switch is closed (open) when $S = 1$ ($S = 0$). An electrical ladder diagram that represents the control of the light using the switch is shown in Figure 8.1. As indicated in the figure, the left rail is the power rail and the right rail is the neutral (or ground) rail. When the switch is closed, there is an electrical path between the left and right rails, and thus power flows through the light so that the light is on when the switch is closed. When the switch is open, the electrical path between the rails is broken and there is no power flow through the light. Hence in this case the light is off.

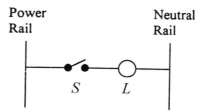

Figure 8.1   Electrical ladder diagram for light with switch.

The ladder diagram in Figure 8.1 has only one "rung" through which power flows to the light when the switch is closed. Note that the operation of the rung is given by the Boolean logic equation

$$L = S;$$

that is, $L = 1$ if and only if $S = 1$.

Multiple rungs are required to control a collection of lights (or various devices) using switches. For example, suppose that we want to control two lights using three switches. The status of Light 1 and Light 2 is denoted by L_1 and L_2, and the status of Switches 1, 2, and 3 is denoted by S_1, S_2, and S_3. A two-rung electrical ladder logic diagram for the interconnection of the lights with the switches is shown Figure 8.2. As seen from the figure, there is an electrical path from the left to the right rail through Light 1 if either Switch 1 or Switch 2 is closed; that is, $S_1 = 1$ or $S_2 = 1$. There is an electrical path from the left to the right rail through Light 2 if both Switch 2 and Switch 3 are closed; that is, $S_2 = 1$ and $S_3 = 1$. Hence, Light 1 is on if either $S_1 = 1$ or $S_2 = 1$, and Light 2 is on if $S_2 = 1$ and $S_3 = 1$. The Boolean logic equations for the two rungs are

$$\text{Rung 1:} \quad L_1 = S_1 + S_2$$

$$\text{Rung 2:} \quad L_2 = S_2 S_3$$

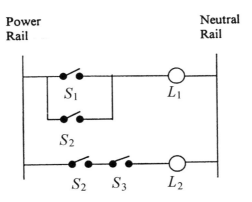

Figure 8.2   Two-rung ladder diagram.

Control Relays

A common component of a hardwired control implementation is a control relay. As illustrated in Figure 8.3, a control relay consists of a coil and normally open (n.o.) and normally closed (n.c.) contacts. When there is no current applied to the coil, the coil is not energized and the normally open contacts remain open and the normally closed contacts remain closed. When current is applied to the

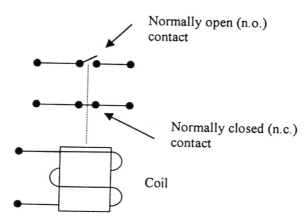

Figure 8.3 Control relay with normally open and normally closed contacts.

coil, the coil is energized, creating an electromagnetic field that closes the normally open contacts and opens the normally closed contacts.

An example of an electrical ladder logic diagram containing a control relay is shown in Figure 8.4. In this ladder diagram, CR, which stands for "control relay," denotes the status of the coil with $CR = 1$ when the coil is energized and $CR = 0$ when the coil is not energized. The normally open contact of the control rely is denoted by the symbol

and the normally closed contact is denoted by the symbol

The status of the normally open contact is given by CR (which denotes the status of the coil), and the status of the normally closed contact is given by CR.

For the control relay in Figure 8.4, the coil is energized when the switch is closed ($S = 1$). As seen from the ladder diagram, the contacts of the relay control Light 1 and Light 2. When the switch is open, the coil is not energized, and thus the normally open contact is open so that Light 1 is off ($L_1 = 0$), and the normally closed contact is closed so that Light 2 is on ($L_2 = 1$). When the switch is closed and the coil is energized, the normally open contact is closed so that Light 1 is on ($L_1 = 1$), and the normally closed contact is open so that

Light 2 is off ($L_2 = 0$). The Boolean logic equations for the rungs in the ladder diagram are

$$\text{Rung 1:} \quad CR = S$$

$$\text{Rung 2:} \quad L_1 = CR$$

$$\text{Rung 3:} \quad L_2 = \overline{CR}$$

Combining the equations for the rungs yields

$$L_1 = S \quad \text{and} \quad L_2 = \overline{S}$$

For an application example, we can use a control relay to implement pushbutton control of a motor: As was first considered in Chapter 7, suppose we want a motor to turn on when the start pushbutton switch is pressed and turn off when the stop pushbutton switch is pressed. We again assume that each switch closes momentarily when pushed and then reopens. We let M denote the status of the motor and $PB1$ and $PB2$ the status of the pushbutton switches, where again 1 means on or closed and 0 means off or open.

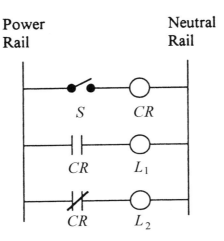

Figure 8.4 Electrical ladder logic diagram with control relay.

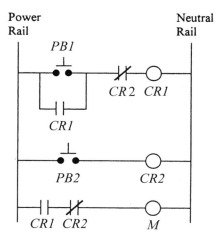

Figure 8.5 Electrical ladder logic diagram for pushbutton control of a motor.

To generate a hardwired implementation of the desired control action, it turns out that two control relays are needed. The electrical ladder logic diagram for the implementation is shown in Figure 8.5. From the diagram we see that when the start pushbutton switch is pressed and the coil of Control Relay 2 is not energized ($CR2 = 0$), the coil of Control Relay 1 is energized ($CR1 = 1$), which closes the relay's normally open contact and turns on the motor. The motor stays on until the stop pushbutton switch is pressed, which energizes the coil of Control Relay 2 so that $CR2 = 1$. The relay's normally closed contact then opens, which turns the motor off. The Boolean logic equations for the rungs in the ladder diagram are

$$\text{Rung 1:}\quad CR1 = (PB1 + CR1)\overline{CR2}$$

$$\text{Rung 2:}\quad CR2 = PB2$$

$$\text{Rung 3:}\quad M = CR1\,\overline{CR2}$$

The equation for Rung 1 shows that Control Relay 1 is "latched" whenever Pushbutton Switch 1 is pushed. This means that if Pushbutton Switch 2 is not pushed ($PB2 = 0$), the coil remains energized ($CR1 = 1$) even though $PB1$ is reset to 0.

Combining the equations for the rungs gives

$$M = (PB1 + CR1)\left(\overline{PB2}\right)\left(\overline{PB2}\right) = (PB1 + CR1)\overline{PB2} \qquad (8.1)$$

From (8.1) we see that the motor is on if the stop pushbutton switch is not pressed ($PB2 = 0$) and either the start pushbutton switch is pressed ($PB1 = 1$) or the coil of Control Relay 1 is energized ($CR1 = 1$). This is the latching operation that was first pointed out in Chapter 7; that is, the motor is latched so that it stays on as long as the stop pushbutton switch is not pressed.

Software Ladder Logic Diagrams

A PLC can be programmed using a "software ladder logic diagram" that is very similar to an electrical ladder logic diagram. In a software ladder logic diagram, input conditions, such as the status of on/off switches, contacts of coils, pushbutton switches, limit switches, etc., are denoted by the contact symbols:

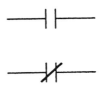

When the upper symbol appears in a software ladder logic diagram, it means that the input is "examined for an on condition," and when the lower symbol appears, it means that the input is "examined for an off condition." To illustrate the use of the "examine on" convention, in Figure 8.6 we give the software ladder logic diagram corresponding to the electrical ladder logic diagram in Figure 8.2. Recall that L_1 and L_2 denote the status of the two lights, and S_1, S_2, and S_3 denote the status of the switches. In this case, all the inputs (the switches) are examined for an "on condition." Hence, Light 1 is on when S_1 is on or S_2 is on, and Light 2 is on when both S_2 and S_3 are on.

Now suppose the software ladder diagram is modified to include two "examine for off" inputs as given in Figure 8.7. In this case, L_1 is on when S_1 is off or S_2 is on, and L_2 is on when S_2 is on and S_3 is off.

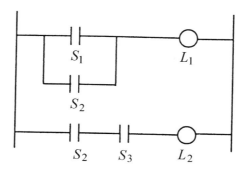

Figure 8.6   Software ladder logic diagram for the control of two lights.

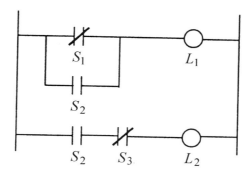

Figure 8.7   Modified software ladder logic diagram.

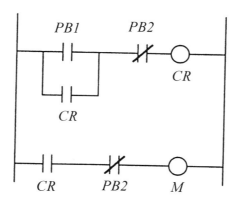

Figure 8.8   Software ladder logic diagram for motor with pushbutton switches.

The software version of a control relay is a software relay with normally open or closed contacts. For example, the software version of the control relay implementation of the motor with two pushbutton switches (see Figure 8.5) is shown in Figure 8.8. Note that only one software relay is needed in the ladder diagram in Figure 8.8, as opposed to the two control relays that were required in the electrical ladder diagram in Figure 8.5. The reason for this is that in the software version, *PB2* can be examined for the off condition, and thus the normally closed contact of a control relay is not needed. However, we still need a software relay to achieve the latching condition.

Ladder Diagrams from Boolean Logic Equations

Any *p*-input *m*-output *N*-state-variable discrete logic controller given by Boolean logic equations can be specified in terms of a software ladder logic diagram. The ladder diagram can be directly constructed from the Boolean logic equations as follows.

Suppose that the state variables of the controller are denoted by X_1, X_2, ..., X_N. In Chapter 7 it was shown that if Set_{X_i} and Reset_{X_i} cannot both be 1 at the same time or if X_i^+ can be set equal to 0 when $\text{Set}_{X_i} = \text{Reset}_{X_i} = 1$ and $X_i = 0$, the state equations can be expressed in the standard reset/set flip flop form

$$X_i^+ = (\overline{\text{Reset}_{X_i}})\left[X_i + \text{Set}_{X_i}\right], \quad i = 1, 2, ..., N \qquad (8.2)$$

where the Set_{X_i} and Reset_{X_i} operations are functions of the input variables I_1, I_2, ..., I_p and the other state variables. Also recall from Chapter 7 that the outputs Y_1, Y_2, ..., Y_m are given by

$$Y_i = h_i(X_1, X_2, ..., X_N), \quad i = 1, 2, ..., m \qquad (8.3)$$

The general form of the software ladder logic diagram of the controller given by (8.2) and (8.3) is shown in Figure 8.9. The detailed ladder diagram for a given system can be computed from the diagram in Figure 8.9 by expressing the Set_{X_i} and Reset_{X_i} operations and the functions $h_i(X_1, X_2, ..., X_N)$ in terms of contacts for the inputs and state variables.

In the ladder diagram in Figure 8.9, note that the status X_i of the software coils (given by the circles) is the next value of the state variable X_i. The superscript "+" notation for the next state has been dropped in the ladder diagram

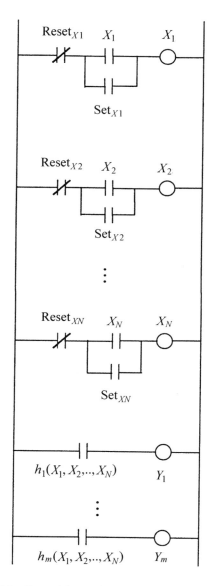

Figure 8.9 General form of software ladder logic diagram.

since it is understood that the coil value is the next state value. It is important to point out that in the evaluation of the ladder logic diagram in Figure 8.9, it is assumed that the state values are not updated until all the new state values have been computed. The case when the diagram is evaluated sequentially (that is, from the first to the last rung,) is discussed later.

To illustrate the generation of the software ladder logic diagram, let us again consider the level controller that was studied in Chapter 7. In this example [see (7.31) and (7.32)], the state equations are given by

$$X_1^+ = (\overline{LLS})(X_1 + X_2)$$

$$X_2^+ = \overline{LLS}$$

where X_1 is the status of the pump, X_2 is the status of the input valve, and *LLS* is the status of the level limit switch. The software ladder logic diagram for the controller is shown in Figure 8.10. In this case, since the outputs are given by $Y_1 = X_1$ and $Y_2 = X_2$, there is no need to include rungs for the outputs in the ladder diagram.

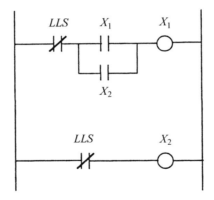

Figure 8.10 Software ladder logic diagram of level controller.

PLC Implementation

A block diagram of a PLC implementation of a discrete logic controller is shown in Figure 8.11. The inputs $I_1, I_2, \ldots, I_p$ are signals from the input devices (switches, etc.) that represent the on/off status of the input devices, and the ouputs Y_1, Y_2, ..., Y_m are the commands that are sent to the output devices to control their on/off status. The inputs are applied to an I/O interface that puts the signals into the proper format for processing by the central processing unit (CPU). The output of the CPU then goes through another I/O interface that converts the CPU output into command signals that are sent to the output devices.

The status of the inputs and outputs are stored in the I/O interfaces using registers that have numerical addresses for reference purposes. There are also registers in the CPU that store the status of the software coils comprising the PLC control program. The CPU reads the status of the inputs stored in the I/O registers, executes the PLC program, and then updates the registers in the I/O interface that determine the command signals that are sent to the output devices. The process of reading the status of the input devices, executing the program, and updating the output commands is refered to as a *scan*. The PLC repeats scans on a periodic basis with a fixed time for a scan. The time for a scan depends on the PLC and can vary from several milliseconds to a few microseconds.

The PLC program is generated by "inputting" a software ladder logic diagram into the PLC with numerical addresses selected for the input devices, software coils, and output devices. The addressing of registers is defined by displaying a representation of the I/O modules along with the software ladder logic diagram. For example, the I/O modules and ladder diagram for the PLC implementation of the level controller (see Figure 8.10) are shown in Figure 8.12. The PLC program in Figure 8.12 is executed by evaluating the rungs in sequential order beginning with the first rung. That is, the new value of the input is extracted from address 30, then the new value for the coil with address 40 is evaluated, and next the new value for the coil with address 41 is evaluated. Note that the coil with address 40 corresponds to the state variable X_1, and the the coil with address 41 corresponds to the state variable X_2.

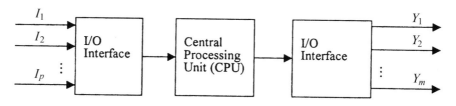

Figure 8.11   Block diagram of a PLC implementation

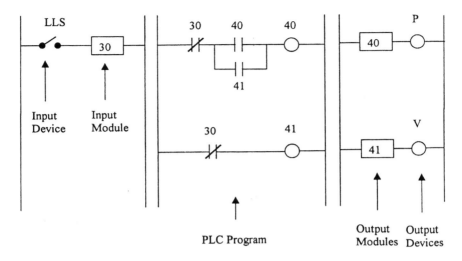

Figure 8.12 PLC implementation of level controller.

In this example it is important to point out that the value of X_2 is independent of the value of X_1, and thus the next value for X_1 can be computed first, and then the next value for X_2 can be computed. Therefore, the sequential evaluation (rung-by-rung) of the ladder diagram yields the correct result in this case. However, in the general case when the state equations are given by (8.2), the value of the ith state variable X_i may depend on the values of the state variables $X_1, X_2, ..., X_{i-1}$. Hence if the values of $X_1, X_2, ..., X_{i-1}$ are updated first, and then the value of X_i is updated using the updated values of $X_1, X_2, ..., X_{i-1}$, the result may not be the same as in the case when X_i is updated using the previous values of $X_1, X_2, ..., X_{i-1}$.

To allow for the sequential evaluation of the state variables, one can attempt to define the state variables $X_1, X_2, ..., X_N$ so that for any $i = 2, 3, ..., N$, the value of X_i is independent of the values of $X_1, X_2, ..., X_{i-1}$. If this is not possible, one can simply add additional state variables. For example, consider the $N = 2$ case given by the state equations

$$X_1^+ = f_1(X_1, X_2, I) \tag{8.4}$$

$$X_2^+ = f_2(X_1, X_2, I) \tag{8.5}$$

$$Y = h(X_1, X_2) \tag{8.6}$$

In general, the value of X_2 depends on the value of X_1, and thus if the updated value of X_1 is used in (8.5) to compute X_2^+, the result may differ from the correct value. In other words, in order for (8.4) and (8.5) to be evaluated sequentially, it must be true that

$$X_2^+ = f_2(X_1^+, X_2, I) \tag{8.7}$$

To eliminate this problem, we define a new state variable X_3 with

$$X_3^+ = X_1 \tag{8.8}$$

and we modify (8.5) so that

$$X_2^+ = f_2(X_3, X_2, I) \tag{8.9}$$

This ensures that the value of X_1 used to compute X_2^+ is the previous value of X_1. Then expressing (8.4) and (8.8) in the reset/set flip flop form we obtain the software ladder diagram for the general two state variable ($N = 2$) case shown in Figure 8.13.

$$X_1^+ = (\overline{\text{Reset}_{X_1}})\left[X_1 + \text{Set}_{X_1}\right]$$

$$X_2^+ = (\overline{\text{Reset}_{X_2}})\left[X_2 + \text{Set}_{X_2}\right]$$

To summarize, the generation of a discrete logic controller can be carried out by using the state variable approach with a PLC implementation of the resulting Boolean logic equations. The major steps of the design procedure are as follows:

1. Define appropriate state variables and determine the state transition diagram for each of the state variables.

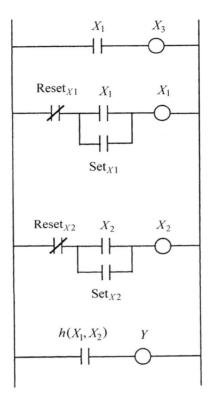

Figure 8.13 Software ladder logic diagram in $N = 2$ case.

2. Generate Boolean logic equations from the state transition diagrams using the set/reset operations.
3. Generate a software ladder logic diagram from the Boolean logic equations.
4. Generate a PLC program from the software ladder logic diagram by selecting appropriate addresses for the inputs, software coils, and outputs.

We conclude this chapter by giving an example of the design procedure.

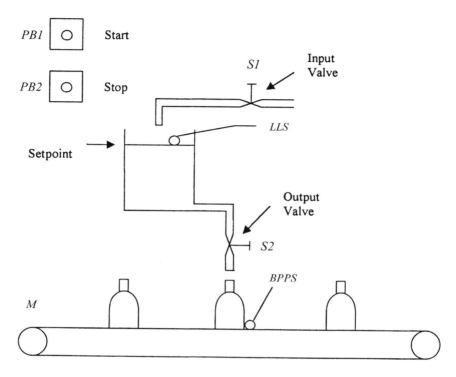

Figure 8.14   Bottle-filling operation.

Bottle-Filling Operation

In this example the objective is to design a discrete logic controller for the bottle-filling operation illustrated in Figure 8.14. The input devices to the controller are the start pushbutton switch, stop pushbutton switch, level limit switch, and bottle position proximity switch. The on/off status of these devices is denoted by *PB1*, *PB2*, *LLS*, and *BPPS*, respectively. The output devices to be controlled are a solenoid that drives the input valve, a solenoid that drives the output valve, and the conveyor motor. The on/off status of the output devices is denoted by *S1*, *S2*, and *M*, respectively. The desired control action is given by:

1.   The process is turned on by pressing the start pushbutton button switch (*PB1* = 1).

2. The process is turned off by pressing the stop pushbutton switch (*PB2* = 1), where "off" means that the solenoids are deenergized and the conveyor motor is off.
3. When the process is operating:
   - The solenoid that drives the input valve is energized when the level in the tank is less than the set point, in which case *LLS* = 0.
   - After a bottle arrives in position, the solenoid that drives the output valve is energized and remains energized for 10 seconds which is the time required to fill a bottle.
   - After a bottle is filled, the solenoid that drives the output valve is deenergized and the conveyor motors turns on.
   - The conveyor is on when a bottle is not in position (*BPPS* = 0) and is off when a bottle is in position (*BPPS* = 1) and is being filled.
   - The filling of bottles is then repeated over and over again.

To design the discrete logic controller, we first define the state variables $X_1 = S_1$, $X_2 = S_2$, and $X_3 = M$. In this example, it turns out that we need two additional state variables: one for the timer coil needed to realize the 10-sec bottle-filling time and one for process on or off. Hence, we need to define

$$X_4 = \text{timer coil on/timer coil off}$$

$$X_5 = \text{process on/ process off}$$

In the PLC implementation of the timer coil, we use a TON software timer, where TON stands for "time-to-on." In a TON timer, the normally open contacts (resp., normally closed contacts) close (resp., open) T seconds after the timer coil is energized, which in this example is 10 sec (i.e., $T = 10$ sec).

From the specification of the control action given previously, we can determine the state transition diagram for each state variable. A description of the state transitions for each state variable is follows:

- X_1 is set from 0 to 1 when the process is on ($X_5 = 1$) and the level in the tank is less than the set point (*LLS* = 0). X_1 is reset to 0 from 1 when the process is turned off ($X_5 = 0$) or the level in the tank is at or above the set point (*LLS* = 1).

- X_2 is set to 1 from 0 when the process is on and a bottle is in position for filling ($BPPS = 1$). X_2 is reset to 0 from 1 when the process is turned off or a normally closed contact of the timer coil opens $[(X_4)_{n.c.} = 0]$.
- X_3 is set to 1 from 0 when the process is on and a bottle is not in position for filling or the process is on, a bottle is in the filling position, and a normally open contact of the timer coil closes $[(X_4)_{n.o.} = 1]$. [Note that $(X_4)_{n.o.} = 1$ is the same as $(X_4)_{n.c.} = 0$.] X_3 is reset to 0 from 1 when the process is turned off or a bottle is in the filling position.
- X_4 is set to 1 from 0 when the process is on and a bottle is in position for filling. X_4 is reset to 0 from 1 when the process is turned off or a bottle is not in position for filling.
- X_5 is set to 1 from 0 when the start pushbutton switch is pressed ($PB1 = 1$). X_5 is reset to 0 from 1 when the stop pushbutton switch is pressed ($PB2 = 1$).

The state transition diagrams are shown in Figure 8.15. From the figure, we have

$$\text{Set}_{X_1} = X_5\overline{LLS}, \quad \text{Reset}_{X_1} = \overline{X_5} + LLS$$

$$\text{Set}_{X_2} = X_5 BPPS, \quad \text{Reset}_{X_2} = \overline{X_5} + \overline{(X_4)_{n.c.}}$$

$$\text{Set}_{X_3} = X_5\overline{BPPS} + X_5 BPPS(X_4)_{n.o.}, \quad \text{Reset}_{X_3} = \overline{X_5} + BPPS$$

$$\text{Set}_{X_4} = X_5 BPPS, \quad \text{Reset}_{X_4} = \overline{X_5} + \overline{BPPS}$$

$$\text{Set}_{X_5} = PB1, \quad \text{Reset}_{X_5} = PB2$$

For state variables X_1, X_4, and X_5, Set_X and Reset_X cannot both be equal to 1 at the same time, and for state variable X_2, we can take $X_2^+ = 0$ when $\text{Set}_{X_2} = \text{Reset}_{X_2} = 1$ and $X_2 = 0$. Thus, in these cases, the reset/set equation (8.2) applies; that is,

$$X_i^+ = (\overline{\text{Reset}_{X_i}})\left[X_i + \text{Set}_{X_i}\right] \quad \text{for } i = 1, 2, 4, 5 \quad (8.10)$$

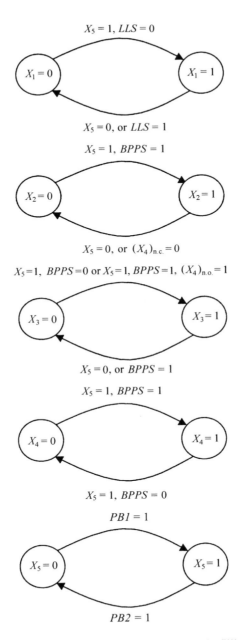

Figure 8.15 State transition diagrams for the bottle-filling operation.

Inserting the reset/set expressions into (8.10) gives

$$X_1^+ = X_5\overline{LLS}\left[X_1 + X_5\overline{LLS}\right]$$

$$X_2^+ = X_5(X_4)_{n.c.}\left[X_2 + X_5BPPS\right]$$

$$X_4^+ = X_5BPPS\left[X_4 + X_5BPPS\right]$$

$$X_5 = \overline{PB2}\left[X_5 + PB1\right]$$

Simplifying these equations yields

$$X_1^+ = X_5\overline{LLS} \tag{8.11}$$

$$X_2^+ = X_5(X_4)_{n.c.}\left[X_2 + BPPS\right] \tag{8.12}$$

$$X_4^+ = X_5BPPS \tag{8.13}$$

$$X_5^+ = \overline{PB2}\left[X_5 + PB1\right] \tag{8.14}$$

Putting (8.11)–(8.14) in words, we have

- The solenoid that drives the input valve is energized if the process is on and the level in the tank is less than the set point; otherwise, the solenoid is off.
- The solenoid that drives the output valve is energized when the process is on, a normally closed contact of the timer coil is closed, and either the solenoid is already energized or a bottle is in position for filling. Otherwise, the solenoid is off.
- The timer coil is energized when the process is on and a bottle is in position for filling; otherwise the timer coil is not energized.
- The process is on when the stop pushbutton is not pressed and either the process is already on or the start pushbutton is pressed; otherwise the process is off.

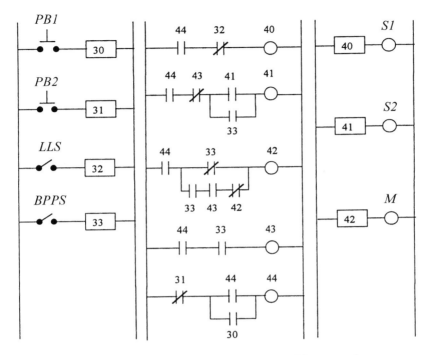

Figure 8.16 PLC implementation of bottle-filling operation.

For state variable X_3, we cannot have $X_3^+ = 0$ when $\text{Set}_{X_3} = \text{Reset}_{X_3} = 1$ and $X_3 = 0$, thus in this case, we must use the state equation having the form (7.22); that is,

$$X_3^+ = \left(\overline{\text{Reset}_{X_3}}\right)X_3 + \left(\text{Set}_{X_3}\right)\overline{X_3}$$

Inserting the expressions for Set_{X_3} and Reset_{X_3} gives

$$X_3^+ = X_5(\overline{BPPS})X_3 + \left[X_5\overline{BPPS} + X_5 BPPS(X_4)_{\text{n.o.}}\right]\overline{X_3}$$

This simplies to

$$X_3^+ = X_5\left[\overline{BPPS} + BPPS(X_4)_{\text{n.o.}}\overline{X_3}\right] \tag{8.15}$$

From (8.15) we have that the conveyor motor is on when the process is on and either a bottle is not in position for filling or a bottle is in position for filling, a normally open contact of the timer coil is closed, and the conveyor motor is not already on. Otherwise, the conveyor motor is off.

From (8.11) – (8.15) we see that X_i^+ does not depend on X_1, X_2, ..., X_{i-1}, and thus (8.11) – (8.15) can be implemented sequentially using a PLC. Note that if the on/off process state variable had been defined to be X_1 (instead of X_5), all of the other state variables would depend on X_1. In this case the sequential evaluation of the state equations would not correspond to the state equations (8.11) – (8.15). We can generate a PLC program for the the bottle-filling operation directly from (8.11) – (8.15). The result is given in Figure 8.16.

Problems

8.1   Give the I/O modules and ladder logic diagram for a PLC implementation of the oven controller in Problem 7.2.

8.2   Give the I/O modules and ladder logic diagram for a PLC implementation of the mixing tank controller in Problem 7.3.

8.3   Give the I/O modules and ladder logic diagram for a PLC implementation of the discrete logic controller in Problem 7.6.

8.4   Give the I/O modules and ladder logic diagram for a PLC implementation of the discrete logic controller in Problem 7.7.

8.5   In the bottle-filling process described in this chapter, suppose that the filling of a bottle is detected using a photosensor rather than using a timing coil. Denote the status of the photosensor by PS, where $PS = 0$ means the bottle is not full and $PS = 1$ means the bottle is full. Modify the I/O modules and ladder diagram in Figure 8.16 so that the timing coil is removed and the photosensor is added.

8.6   A mixing tank has one input valve controlled by a solenoid with status $S1$ through which ingredient A is added. It has a second input valve controlled by a solenoid with status $S2$ through which ingredient B is added. Give the I/O modules and ladder logic diagram for a PLC implementation of a discrete logic controller that carries out the following operations: When the start pushbutton switch with status PB is pressed, ingredient A is added to the tank until a level limit switch opens ($LLS1 = 0$). Then ingredient B is added until a second level limit switch opens ($LLS2 = 0$), at which time the temperature in the tank is raised to $300°$ using a heater with status H. A temperature sensor with status T indicates when the temperature is above or below a set point. Also, after ingredient B is added, a mixer motor with status M is turned on, and stays on for 20 minutes during which time the temperature in the tank is maintained at $300°$. After 20 min, a drain valve controlled by a solenoid with status $S3$ is opened, and then is closed when the empty tank switch opens ($ETS = 0$) indicating that the tank is empty. The process is then repeated by pressing the start pushbutton switch.

Chapter 9

Manufacturing Systems

As first noted in Chapter 1, a manufacturing operation can be viewed as a system with inputs equal to the raw materials and with outputs equal to finished materials or products. This input/output view of a manufacturing system is illustrated in Figure 9.1. A manufacturing system is specified in terms of a collection of stations (machines, processes, or work centers) that are required to produce the product. The equipment needed to manufacture the product comprises the machine or process level of manufacturing hierarchy. As noted in Chapter 1, this is the lowest level in the manufacturing hierarchy. The operation of the equipment in a manufacturing system often involves both continuous-variable time-driven controllers and discrete logic controllers. Hence, the control methods discussed in Chapters 4 through 8 are directly applicable to the operation of manufacturing systems.

Above the machine/process level in the manufacturing hierarchy is the production level. The production level consists of the cell/line level, which contains groups of machines, processes, or workstations, and the factory floor level, which contains the complete production facility. A key aspect of production is the movement or routing of materials, parts, or assemblies (called *jobs*) through the manufacturing facility, and the processing of jobs at the individual stations. There are two basic formats for organizing the operation of a manufacturing system: *job shop* and *flow line*. In a job shop, jobs may enter the manufacturing facility at any particular station, and then may have different routings from station to station, which may include visiting the same station more than once. In a flow line, also called a *flow shop*, all jobs enter at the same station and then are processed station-by-station in a fixed sequence. In contrast to a job shop, jobs in a flow line visit a station once and only once. Job shops are used primarily for low-volume customized products, whereas flow lines are used primarily for high-volume mass production. Here "volume" refers to the amount of product being produced.

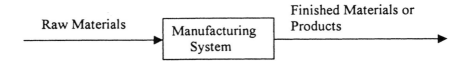

Figure 9.1 Input/output representation of a manufacturing system.

There are two basic types of production activities:

1. *Value-adding operations* on materials and parts such as machining, assembly, shape alteration, material deposition, mixing, heating/ cooling, etc.
2. *Nonvalue-adding operations* such as materials handling, waiting for processing, temporary storage, and test and/or inspection

The ratio of the time required for the value-adding operations and the time required for the nonvalue-adding operations is a very important factor in determining the efficiency (and hence cost) in manufacturing products. We consider this next along with other performance measures for evaluating a manufacturing system.

Performance Measures

Performance measures are very useful in characterizing the behavior of a manufacturing system. They are particularly important in determining control requirements on both the machine/process level and the production level. To define various measures, suppose that the objective is to manufacture Q quantities of a product, where Q is called the *lot size* or the *batch size*. It is assumed that the manufacturing of the product requires that a part be processed at N machines or stations, and the sequence of processing is given by

$$M_1 \rightarrow M_2 \rightarrow \cdots \rightarrow M_N \qquad (9.1)$$

where M_i denotes the ith machine. It is possible that in going through sequence (9.1), the parts may visit the same machine more than once, in which case M_i

and M_j in the sequence (9.1) may denote the same machine. It is assumed that the batch of Q quantities is moved together in some container from machine to machine, and thus buffers (temporary storage locations) are needed at each machine. This type of processing is usually cheaper than having to move each part in the batch one-by-one, especially when the parts are small in size.

The *manufacturing lead time* (MLT) is the amount of time required to manufacture the batch of Q quantities. The MLT specifies when the manufacturing process should be started in order to be able to meet the delivery date (due date) specified by the customer. To derive a formula for the MLT, let

T_{oi} = operating or processing time per part on machine M_i,

T_{nvoi} = time for the nonvalue-adding operations per batch associated with machine M_i,

T_{si} = time to set up machine M_i for the batch to be produced.

Here all times are in the same units (usually hours).

With these definitions, the MLT is given by

$$\text{MLT} = \sum_{i=1}^{N}\left[QT_{oi} + T_{nvoi} + T_{si} \right] \qquad (9.2)$$

The *actual processing time* (APT) is given by

$$\text{APT} = \sum_{i=1}^{N} QT_{oi} \qquad (9.3)$$

The ratio of MLT to APT is very important. Ideally, we would like to have MLT/APT = 1, but in practice, MLT may be much larger than APT due to the need for setup time and nonvalue-adding operations.

For an example on the computation of the MLT, suppose that the manufacturing of a product requires that four machines process a part. The values for the times are shown in Table 9.1. With $Q = 100$, evaluation of (9.2) gives

$$\text{MLT} = \left(100\frac{6}{60} + 4 + 4 \right) + \left(100\frac{9}{60} + 5 + 3 \right) + \left(100\frac{7}{60} + 5 + 5 \right) + \left(100\frac{8}{60} + 6 + 6 \right)$$
$$= 18 + 23 + 21.67 + 25.33 = 88 \text{ hr.}$$

Machine number	1	2	3	4
T_{oi} (min)	6	9	7	8
T_{nvoi} (hr)	4	5	5	6
T_{si} (hr)	4	3	5	6

Table 9.1 Times for MLT Computation

Evaluating (9.3) yields

$$\text{APT} = 100\left(\frac{6}{60} + \frac{9}{60} + \frac{7}{60} + \frac{8}{60}\right) = 100\frac{30}{60} = 50 \text{ hr}$$

In this case, MLT/APT = 88/50 = 1.76.

Note that if $T_{oi} = T_o$, $T_{nvoi} = T_{nvo}$, and $T_{si} = T_s$ for all i, so that the times are the same for all machines, (9.2) and (9.3) reduce to

$$\text{MLT} = N\left(QT_o + T_{nvo} + T_s\right) \tag{9.4}$$

$$\text{APT} = NQ(T_o) \tag{9.5}$$

In the following development, it is assumed that $T_{oi} = T_o$, $T_{nvoi} = T_{nvo}$, and $T_{si} = T_s$ for all i so that the MLT is given by (9.4).

Production Rate

The production rate can be defined as follows. First, the *batch production time* (BPT) *per machine* is defined by

$$\text{BPT} = QT_o + T_{nvo} + T_s \tag{9.6}$$

where BPT is given in units of hours per machine. The BPT is the total time to process the batch at each machine. Combining (9.4) and (9.6) gives

$$\text{MLT} = N(\text{BPT}) \tag{9.7}$$

The BPT *per machine per part* is denoted by T_p and is given by

$$T_p = \frac{\text{BPT}}{Q} \qquad (9.8)$$

Note that the units of T_p are hours per machine per part.

The *production rate per machine*, denoted by R_p, is the number of parts processed per machine per hour. The production rate per machine R_p is equal to the inverse of T_p; that is,

$$R_p = \frac{1}{T_p} \qquad (9.9)$$

Inserting (9.8) into (9.9) gives

$$R_p = \frac{Q}{\text{BPT}} \qquad (9.10)$$

and using (9.6), we have

$$R_p = \frac{Q}{QT_o + T_{no} + T_s} \qquad (9.11)$$

Finally, dividing the numerator and denominator in (9.11) by Q yields

$$R_p = \frac{1}{T_o + \dfrac{T_{nvo} + T_s}{Q}} \qquad (9.12)$$

It follows from (9.12) that as the batch size $Q \rightarrow \infty$, $R \rightarrow 1/T_o$. Thus the maximum possible production rate per machine is equal to the inverse of the processing time per part for the machine.

For an example on the calculation of the production rate, suppose that for each machine in a production facility the times are $T_o = 4$ min, $T_{nov} = 6$ hr, and $T_s = 4$ hr. Then from (9.6), if $Q = 100$ the batch production time is

$$\text{BPT} = 100\frac{4}{60} + 6 + 4 = 16.67 \text{ hr per machine}$$

From (9.8) we have

$$T_p = \frac{\text{BPT}}{Q} = \frac{16.67}{100} = 0.1667 \text{ hr per machine per part}$$

and, thus,

$$R_p = \frac{1}{T_p} = \frac{1}{0.1667} = 6 \text{ parts per machine per hour}$$

In this case the maximum possible production rate (as $Q \rightarrow \infty$) is

$$(R_p)_{max} = \frac{1}{T_o} = \frac{1}{4/60} = 15 \text{ parts per machine per hour}$$

Capacity

The capacity C is the total number of parts that can be processed in a week assuming that all machines are operated simultaneously. The capacity is given by

$$C = (\text{number of machines}) \times (\text{total working hours per week}) \times (\text{production rate in parts per machine per hour}).$$

If each part must be processed by N machines to manufacture the product, the number of products that can be produced in a week is equal to C/N assuming steady-state operation (no machine failures and raw materials are always available).

For the preceding example where $T_o = 4$ min, $T_{nov} = 6$ hr, and $T_s = 4$ hr, suppose that 8 machines are required to produce the product and the number of working hours in a week is equal to 40. Then the capacity C is

$$C = (8)(40)(6) = 1920 \text{ parts per week}$$

and

$$\frac{C}{N} = \frac{1920}{8} = 240 \text{ products per week}$$

Work-In-Process (WIP)

The work-in-process (WIP) is the total number of parts in the manufacturing system at a given point in time during production. If the system has N machines that are operated simultaneously, and if each machine is processing a batch of Q parts, then the WIP is

$$\text{WIP} = NQ \qquad (9.13)$$

It turns out that WIP is also given by

$$\text{WIP} = R_p(\text{MLT}) \qquad (9.14)$$

To verify (9.14), insert (9.10) and (9.7) into (9.14). This gives

$$\text{WIP} = \left[\frac{Q}{\text{BPT}}\right] N(\text{BPT}) = QN$$

and thus (9.13) does follow from (9.14).

A major goal is to have WIP be as small as possible, since a large WIP implies that a large amount of work-to-be-processed is in the manufacturing system and is not making money for the company (until the product is manufactured and sold). In addition, the facilities necessary to hold large amounts of material may be expensive to build and maintain.

The expression WIP $= NQ$ for the WIP shows that the smallest value of WIP can be obtained by taking $Q = 1$ (i.e., a batch size of 1). However, taking $Q = 1$ implies that the parts must be moved through the system one-by-one. This could be expensive to carry out, and thus there is a trade-off between the size of WIP and the materials handling requirements.

Flow-Line Analysis

We now focus on flow lines, also called *transfer lines*, that are used in high-volume production such as electronics assembly and automobile parts (e.g., engines). Parts are usually moved through such systems one-by-one, and thus the batch size Q is equal to 1. This is assumed to be the case for the following analysis. Note that the analysis given previously applies to general production systems and not just flow lines; whereas, the analysis given below is applicable only to flow lines.

The processing sequence for a flow line is given by

$$\text{parts in} \ \rightarrow B_1 \rightarrow M_1 \rightarrow B_2 \rightarrow M_2 \rightarrow \cdots \rightarrow B_N \rightarrow M_N \rightarrow \text{parts out}$$

where M_1, M_2, ..., M_N are the machines or stations and B_1, B_2, ..., B_N are the buffers (temporary storage locations). In some cases the buffers are the materials handling equipment between stations, such as conveyors. For example, in electronics assembly there may be several printed circuit boards on the conveyor between the assembly stations.

We shall model the flow line in terms of the input $A(k)$ and the output $C(k)$ where $A(k)$ is equal to the time at which the kth part is put into the system and $C(k)$ is the time at which the line is done with the kth part. It is assumed that parts are put into the line one after another so that $A(k) > A(k-1)$.

If $A(k) = (k-1)\lambda$ where λ is a fixed positive constant, the first part is put in at time $A(1) = 0$, the second part is put in at time $A(2) = \lambda$, the third part is put in at time $A(3) = 2\lambda$, and so on. When $A(k) = (k-1)\lambda$, the time interval between input times is

$$A(k) - A(k-1) = (k-1)\lambda - [(k-2)\lambda] = \lambda$$

Then defining the input rate to be the inverse of the time interval between input times, we have that

$$\text{Input rate} \ = \ \frac{1}{A(k) - A(k-1)} \ = \ \frac{1}{\lambda} \qquad (9.15)$$

Similarly, the output rate is defined by

$$\text{Output rate} \ = \ \frac{1}{C(k) - C(k-1)} \qquad (9.16)$$

The output rate is the *production rate* or *throughput* of the flow line; that is, it is the number of products per unit time that are produced by the line. When the input rate is a constant given by (9.15), in general the output rate given by (9.16) will vary as a function of the part number k, so the output rate may not be constant. Variability of the output rate will result if the availability of raw materials is delayed, the machine processing times are variable, or if the machines occasionally break down.

We can study the behavior of the flow line by attempting to express $C(k)$ as a function of $A(k)$. We begin the analysis with the single-buffer single-machine case.

Single Machine with Buffer

Suppose that the flow line consists of a single machine with a buffer before the machine as illustrated in Figure 9.2. We can express $C(k)$ as a function of $A(k)$: First, it is assumed that the machine with buffer operates on parts using the *first-in first-out* (FIFO) *policy*. This means that the machine processes parts in the order in which they are put into the system. We also make the following assumptions:

1. The machine has already been set up so the setup time is neglected.
2. There are no parts in the system at time $t = 0$.
3. The buffer has infinite capacity; that is, it can hold any number of parts.
4. The machine starts processing a part as soon as a part is available at the machine if the machine is not already busy processing another part.
5. The machine processing time is equal to T_o which includes the move time.
6. A part is removed from the system as soon as the machine is done processing the part.

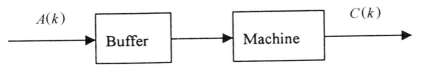

Figure 9.2 Single machine with buffer.

Now we can derive an expression for $C(k)$ as a function of $A(k)$: First, since $A(1)$ is the time at which Part 1 is put into the system and the processing time of the machine is equal to T_o, the machine will finish processing the first part at time $A(1) + T_o$; thus,

$$C(1) = A(1) + T_o \qquad (9.17)$$

The earliest possible time that the machine can finish processing Part 2 is $A(2) + T_o$. But the machine cannot start processing the second part until it is done with the first part, and thus the machine cannot be done with the second part before time $C(1) + T_o$. Hence,

$$C(2) = max[A(2) + T_o, C(1) + T_o]$$

Similarly, the machine cannot be done with Part 3 before time $A(3) + T_o$ and time $C(2) + T_o$; thus,

$$C(3) = max[A(3) + T_o, C(2) + T_o]$$

If we continue, we see that $C(k)$ is given by

$$C(k) = max[A(k) + T_o, C(k-1) + T_o] \quad \text{for } k \ge 2 \qquad (9.18)$$

Now suppose that $A(k) = (k-1)\lambda$ with $\lambda \ge T_o$. Then from (9.17), $C(1) = T_o$, and evaluating (9.18) yields

$$C(2) = max(\lambda + T_o, 2T_o) = \lambda + T_o$$

$$C(3) = max(2\lambda + T_o, \lambda + T_o) = 2\lambda + T_o$$

$$\vdots$$

It is clear that $C(k) = (k-1)\lambda + T_o$ for $k \ge 1$. When $A(k) = (k-1)\lambda$ with $\lambda \le T_o$,

$$C(2) = max(\lambda + T_o, 2T_o) = 2T_o$$

$$C(3) = max(2\lambda + T_o, 2T_o) = 3T_o$$

$$\vdots$$

and thus $C(k) = kT_o$ for $k \geq 1$. From these results we see that the output rate is given by

$$\frac{1}{C(k) - C(k-1)} = \frac{1}{\lambda} \quad \text{when } \lambda \geq T_o \text{ or } \frac{1}{\lambda} \leq \frac{1}{T_o} \qquad (9.19)$$

$$\frac{1}{C(k) - C(k-1)} = \frac{1}{T_o} \quad \text{when } \lambda \leq T_o \text{ or } \frac{1}{\lambda} \geq \frac{1}{T_o} \qquad (9.20)$$

Thus the output rate (the production rate) is equal to the input rate when the input rate is less than or equal to the inverse of the machine processing time T_o, and the output rate is equal to the inverse of the machine processing time when the input rate is greater than or equal to $1/T_o$.

Using these results we can also compute the WIP, which is the number of parts in the system (buffer and machine) at a given point in time. To compute the WIP just before time $C(k)$, let q be the largest positive integer such that $A(q) < C(k)$. Then the WIP just before time $C(k)$ is equal to $q - (k-1) = q - k + 1$, and since one part is in the machine, there are $q - k$ parts in the buffer just before time $C(k)$. For example, suppose that $\lambda = 3$ min and $T_o = 5$ min. Then $A(k) = 3(k-1)$ and $C(k) = 5k$, and setting $k = 10$, the largest integer q such that $A(q) < C(10) = 50$ min is $q = 17$. Hence, just before $C(10) = 50$ min, there are $q - k = 17 - 10 = 7$ parts in the buffer and the WIP is equal to 8. Note that the number of parts in the buffer is growing without bound. This is always the case when the input rate is greater than the inverse of the machine processing time.

Two-Machine Case

Suppose the line consists of two machines with buffers as illustrated in Figure 9.3. Here $A(k)$ is the time the kth part is put into the line, $C_1(k)$ is the time that Machine 1 is done with the kth part, and $C(k)$ is the time that the line is done with the kth part. It is assumed that both machines operate under the FIFO policy and that the processing times of the machines are T_{o1} and T_{o2}, respectively.

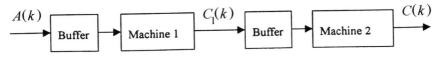

Figure 9.3 Two-machine case.

We assume that the buffers have infinite capacity and the machines process parts as soon as one is available if they are not already busy. Then the equations for the line are

$$C_1(k) = max[C_1(k-1) + T_{o1}, A(k) + T_{o1}] \qquad (9.21)$$

$$C(k) = max[C(k-1) + T_{o2}, C_1(k) + T_{o2}] \qquad (9.22)$$

By solving (9.21) and (9.22) for various values of T_{o1} and T_{o2}, one will see that when $A(k) = (k-1)\lambda$, the steady-state value R_p of the output rate (production rate) is given by

$$R_p = \frac{1}{\lambda} \quad when \quad \frac{1}{\lambda} \le \frac{1}{max(T_{o1}, T_{o2})} \qquad (9.23)$$

$$R_p = \frac{1}{max(T_{o1}, T_{o2})} \quad when \quad \frac{1}{\lambda} \ge \frac{1}{max(T_{o1}, T_{o2})} \qquad (9.24)$$

These results show that the maximum possible production rate is equal to the inverse of $max(T_{o1}, T_{o2})$, and this rate is achievable if and only if the input rate $1/\lambda$ is greater than or equal to $1/max(T_{o1}, T_{o2})$. The machine with the larger processing time, which is equal to the maximum of T_{o1} and T_{o2}, is called the *bottleneck machine* since the fastest rate at which parts can be processed through the line is equal to the inverse of the processing time of the bottleneck machine. If the input rate $1/\lambda$ is greater than $1/max(T_{o1}, T_{o2})$, but is less than or equal to $1/min(T_{o1}, T_{o2})$, the number of parts in the buffer before the bottleneck machine will grow without bound.

General N Machine Case

Now suppose that there are N machines in the line with an infinite-capacity buffer before each machine. If machine M_i has processing time T_{oi}, the bottleneck machine is the one with the largest processing time that is equal to $(T_o)_{max} = max(T_{o1}, T_{o2}, ..., T_{oN})$. The fastest possible production rate is equal to $1/(T_o)_{max}$, and this rate is achievable if and only if the input rate $1/\lambda$ is greater than or equal to $1/(T_o)_{max}$.

The WIP can be computed as was done earlier in the single-machine case: For a given value of k, let q be the largest positive integer such that $A(q) < C(k)$, where again $A(k)$ is the time when the kth part is put into the line and $C(k)$ is the time when the line is done with the kth part. Then the WIP just before time $C(k)$ is equal to $q - (k - 1)$.

If the WIP and the difference $D = C(k) - A(k)$ are constant over some interval of time, the WIP can be computed using *Little's law*, which states that

$$\text{WIP} = R_p D \tag{9.25}$$

In (9.25), R_p is the production rate and $D = C(k) - A(k)$ is the difference between the time a part comes out of the line and the time when the part enters the line. The quantity D is called the *delay*. The smallest possible value for the delay D is equal to the sum of the machine processing times; that is,

$$D = \sum_{i=1}^{N} T_{oi} \tag{9.26}$$

From (9.25), we see that if the WIP is increased while R_p is kept fixed, the delay D increases. This is another reason (in addition to cost issues) for keeping the WIP at a "reasonable" level.

If $R_p = 1/(T_o)_{max}$, (9.25) becomes

$$\text{WIP} = \frac{D}{(T_o)_{max}} \tag{9.27}$$

Inserting (9.26) into (9.27) gives

$$\text{WIP} = \frac{\sum_{i=1}^{N} T_{oi}}{(T_o)_{max}} \tag{9.28}$$

The expression (9.28) for the WIP is called the *critical WIP*, or the *ideal WIP*, since it is the value of the WIP assuming that the maximum possible production rate is achieved with the smallest possible delay. Note that when all the machine processing times $T_{o1}, T_{o2}, \ldots, T_{oN}$ are equal, the critical WIP is equal to N, the number of machines in the line.

In general the critical WIP is not equal to N. For example, suppose that there are 10 machines in the line with processing times equal to 3, 3, 4, 5, 5, 6, 6, 6, 7, and 8 min. Then (9.28) becomes

$$\text{WIP} = \frac{3+3+4+5+5+6+6+6+7+8}{8} = 6.627$$

Rounding this off to the nearest integer gives a critical WIP of 7.

Flow-Line Analysis With Machine Breakdowns

We continue to assume that the line contains N machines with infinite-capacity buffers before the machines. The machines have processing times T_{o1}, T_{o2}, ..., T_{oN} with $(T_o)_{max} = max(T_{o1}, T_{o2}, ..., T_{oN})$. Recall that the fastest possible production rate is equal to $1/(T_o)_{max}$. The quantity $(T_o)_{max}$ is referred to as the *ideal cycle time* since it is equal to the inverse of the maximum production rate assuming that none of the machines break down. Here the term "cycle" refers to the time interval from the start of processing of a part at a machine to the start of processing of the next part at the same machine.

Let F_i denote the number of stops (breakdowns) per cycle at machine M_i, and let $(T_d)_i$ denote the average downtime in minutes per stop to diagnose a malfunction and fix it. Then with breakdowns included, the average processing time T_{pi} over one cycle for machine M_i is given by

$$T_{pi} = T_{oi} + F_i(T_d)_i \qquad (9.29)$$

The bottleneck machine is now the one with the largest average processing time that is equal to $(T_p)_{max} = max(T_{p1}, T_{p2}, ..., T_{pN})$. The average production rate R_p of the line is then equal to $1/(T_p)_{max}$.

The performance of the line is often given in terms of the *line efficiency* E_∞, which is defined by

$$E_\infty = \frac{(T_o)_{max}}{(T_p)_{max}} = \frac{1/(T_p)_{max}}{1/(T_o)_{max}} \qquad (9.30)$$

From (9.30) we see that E_∞ is the ratio of the production rate with breakdowns included to the production rate with no breakdowns. Thus, E_∞ is a measure of the slowdown in production as a result of breakdowns. Clearly, we would like to have E_∞ be as close to 1 (or 100%) as possible.

For an example on the computation of E_∞, suppose that the line has three machines with processing times $T_{o1} = 0.7$ min, $T_{o2} = 0.8$ min, and $T_{o3} = 1.1$

min, so that $(T_o)_{max} = 1.1$ min. The frequency of breakdowns is 0.03 for M_1, 0.05 for M_2, and 0.01 for M_3, and the average downtime $(T_d)_i$ is equal to 10 min for $i = 1, 2, 3$. Then the processing times with breakdowns included are

$$T_{p1} = 0.7 + (0.03)(10) = 1.0$$

$$T_{p2} = 0.8 + (0.05)(10) = 1.3$$

$$T_{p3} = 1.1 + (0.01)(10) = 1.2$$

Note that when the breakdowns are included, the bottleneck machine is M_2; whereas, M_3 is the bottleneck machine if breakdowns are not taken into consideration. Thus, $(T_p)_{max} = 1.3$ min and $R_p = 1/1.3 = 0.77$ products per minute = 46.15 products per hour, and using (9.30) yields

$$E_\infty = \frac{1.1}{1.3} = 0.846 \text{ or } 84.6\%$$

No-Buffer Case

We now assume that there are no buffers before the machines. In this case if any one of the machines goes down, the entire line will stop operation. In particular, if M_i breaks down there is no buffer before M_i that can store the output of M_{i-1}, and thus M_{i-1} must stop operation. The machine M_{i-1} is said to be *blocked* since it cannot pass its output on to the next machine. If M_i breaks down, it is also true that M_{i+1} will not have any parts to process and thus it will stop working. The machine M_{i+1} is said to be *starved*.

To determine the production rate in the no-buffer case we must compute the frequency F that one out of the N machines breaks down. Since machine breakdowns are independent events, the probability that one out of the N machines breaks down is the sum of the probabilities that each machine breaks down. Hence, we can take

$$F = \sum_{i=1}^{N} F_i \qquad (9.31)$$

where F_i is the frequency of breakdown of M_i. If we assume that the time required to fix a breakdown is T_d, the average cycle time T_p at each machine is given by

$$T_p = (T_o)_{max} + FT_d \qquad (9.32)$$

and the average production rate R_p is equal to $1/T_p$.

In the no-buffer case the line efficiency is given by

$$E_o = \frac{(T_o)_{max}}{T_p} \qquad (9.33)$$

We shall compute E_o for the above example where $(T_o)_{max} = 1.1$ min and $F_1 = 0.03$, $F_2 = 0.05$, and $F_3 = 0.1$. Using (9.31) we have

$$F = 0.03 + 0.05 + 0.01 = 0.09$$

and using (9.32) and (9.33) gives

$$R_p = \frac{1}{1.1 + (.09)(10)} = 0.5 \text{ products per minute} = 30 \text{ products per hour}$$

$$E_o = \frac{1.1}{1.1 + (.09)(10)} = 0.55 \text{ or } 55\%$$

Note that the production rate and the efficiency are much less than in the case when there are infinite-capacity buffers before each machine. Thus the use of buffers can greatly enhance performance.

Optimal Location of Buffers

The use of buffers can be expensive and in general buffers increase the WIP, which as noted before is not desirable. If the flow line contains N machines but there are only $M < N$ buffers that can be placed in the line, an interesting problem is that of determining the optimal location for the buffers in order to maximize the line efficiency. A brief treatment of this is given next.

If we assume that the buffers have a sufficiently large capacity, the inclusion of a buffer in the line will "decouple" all the machines before the buffer from all the machines after the buffer. This means that the line efficiency is equal to the minimum of the efficiencies of the two line segments (before and after the buffer). To illustrate this, again consider the example given earlier where $T_{o1} = 0.7$ min, $T_{o2} = 0.8$ min, $T_{o3} = 1.1$ min and $F_1 = 0.03$, $F_2 = 0.05$, and $F_3 = 0.1$. If a buffer is placed before the second machine, the line segments into two parts with the cycle time for the first part equal to $0.7 + (.03)(10) = 1.0$ min, and the cycle time for the second part equal to $1.1 + (0.05 + 0.01)(10) = 1.7$ min. Hence, the production rate is $1/max(1.0, 1.7) = 0.588$ products per minute and the line efficiency E is

$$E = \frac{1.1}{1.1 + 0.6} = 0.647 \text{ or } 64.7\%$$

If a buffer is placed before the third machine, the line segments into two parts with the cycle time for the first part equal to $0.8 + (0.03 + 0.05)(10) = 1.6$ min, and the cycle time for the second part equal to $1.1 + (0.01)(10) = 1.2$ min. Hence, the production rate is $1/max(1.2,1.6) = 0.625$ products per minute and the line efficiency E is

$$E = \frac{1.1}{0.8 + 0.8} = 0.6875 \text{ or } 68.75\%$$

Obviously the production rate and efficiency are higher when the buffer is placed before the third machine, and thus if there is only one buffer that can be placed in the line, it should go before M_3.

Line Balancing

A flow line is said to be *balanced* when the processing time at each station (which could be a single machine) is approximately the same. In a balanced line there is no idle time at the stations, which obviously results in the most efficient mode of operation. If the stations have different processing times, then as seen from the analysis given earlier, the slowest station determines the production rate of the line. To achieve line balancing, it is necessary to arrange the processing steps so that the time at each station is the same. This may require that multiple processing steps be performed at some stations so that the sum of the processing times is the same for all stations. Line balancing is a difficult problem in general since there are always constraints on how the processing steps can be arranged on workstations. In particular, there usually are precedence con-straints that specify which steps can be performed before other steps.

To approach the line balancing problem, we begin by dividing the manu-facturing task into *basic work elements*, which are the smallest indivisible opera-tions comprising the manufacturing task. Let T_i for $i = 1, 2, ..., r$ denote the time required to perform the ith basic work element, where r is the total number of elements. It is assumed that the T_i are constants, although in practice they can vary during production. It is also assumed that the T_i are additive; that is, the time required to perform two work elements is equal to the sum of the times required to perform the individual elements.

Given the values T_i for $i = 1, 2, ..., r$, we define the total work content T_{wc} by

$$T_{wc} = \sum_{i=1}^{r} T_i$$

Then given some desired ideal cycle time T_c for each station, the problem is to arrange the processing of the basic work elements on the stations so that the processing time on each station is equal to T_c and $NT_c = T_{wc}$ where N is the number of stations. It is important to note that the number N of workstations is not known in advance and must be computed as part of the balancing procedure.

The desired ideal cycle time T_c can be determined from the desired produc-tion rate R_p and the expected line efficiency E. From the results given earlier,

$$R_p = \frac{1}{T_p} \quad \text{and} \quad E = \frac{T_c}{T_p}$$

where T_p is the processing time of the bottleneck station. Then inserting $T_p = 1/R_p$ into $E = T_c/T_p$ and solving for T_c yields

$$T_c = \frac{E}{R_p}$$

Thus, to meet the desired production rate, we must have

$$T_c \leq \frac{E}{R_p} \qquad (9.34)$$

The desired ideal cycle time T_c can then be determined from R_p and E by using (9.34).

Given T_c, several methods are available for determining the arrangement of the processing of work elements on stations, but there is no method that gives an optimal solution in terms of maximizing the line balance. All the existing methods yield solutions that may or may not be close to an optimal solution. We consider only the *largest candidate rule* that is specified by the following steps:

1. Arrange the basic work elements in descending order of the processing time T_i with the largest value at the top of the list.
2. Assign work elements to the first station by starting at the top of the list and going down the list, selecting the first appropriate element where appropriate means the element whose processing time is less than or equal to T_c and which satisfies the precedence constraints.
3. Go back to the top of the list and work down, again selecting the next appropriate element for which the sum of the processing time of the previously selected element and the processing time of the next element is less than or equal to T_c and which satisfies the precedence constraints.
4. Repeat Step 3 until no additional elements can be added to the first station without exceeding T_c.
5. Repeat Steps 2 to 4 for the other stations in the line until all of the basic work elements have been assigned to stations in the line.

A measure of the "goodness" of the solution is given in terms of the *balance delay d*, which is defined by

$$d = \frac{NT_c - T_{wc}}{NT_c} \qquad (9.35)$$

The goal is to have d be as small as possible since the larger d is the less balanced the line is; that is, the larger d is the greater the idle time at some stations in the line. Note that $d = 0$ if and only if

$$NT_c = T_{wc} \qquad\qquad (9.36)$$

It follows from (9.36) that the smallest possible number of stations in the line is equal to the smallest integer N that is greater than or equal to T_{wc}/T_c. However, it may not be possible to manufacture the product with this number of stations (unless $d = 0$ can be achieved).

Element number	Processing time (min)	Must be preceded by
1	0.15	—
2	0.4	—
3	0.7	1
4	0.1	1,2
5	0.3	2
6	0.2	3
7	0.6	3,4
8	0.3	5

Table 9.2 Work Elements for Example

To illustrate this procedure, suppose that the manufacturing of a product requires the basic work elements shown in Table 9.2. The precedence constraints are specified by the last column in the table. The "dash" for the first two work elements in the table means that there are no constraints on when these elements can be processed.

From Table 9.2 we see that the longest processing time of the basic work elements is 0.7 min, and thus the smallest possible value of the ideal cycle time T_c is 0.7 min. The fastest possible production rate is then equal to $1/0.7 = 1.43$ products per minute.

From Table 9.2 the total work content is

$$T_{wc} = 0.15 + 0.4 + 0.7 + 0.1 + 0.3 + 0.2 + 0.6 + 0.3 = 2.75 \text{ min}$$

Then with $T_c = 0.7$, $T_{wc}/T_c = 2.75/0.7 = 3.93$, and thus at least 4 stations are required to achieve the fastest possible production rate. However, it turns out that it is not possible to achieve this production rate with 4 stations.

Suppose that we want a production rate of 0.9 products per minute and the expected line efficiency is 0.9 or 90%. Then from (9.34) T_c must be less than or equal to 0.9/0.9 = 1.0 min. We take T_c to be 1.0 min. Then $T_{wc}/T_c = 2.75/1.0$ = 2.75, and thus at least 3 stations are required to manufacture the product. The number of stations that are required and the operations on each station is determined from the largest candidate rule as follows.

First, we list the basic work elements in descending order of the processing times. The result is shown in Table 9.3. Starting at the top of the list in Table 9.3 and going down the list, we see that the first appropriate element is Element 2. Going back to the top of the list and going down, we have that the next appropriate element is Element 5. Continuing, the next element is Element 1 and then Element 4. Again starting at the top of the list and going down, we see that for any remaining elements that can be chosen, the sum of the processing times exceeds 1.0 min. Hence, the work elements to be processed at the first station are Elements 2, 5, 1, and 4.

Element number	Processing time (min)	Must be preceded by
3	0.7	1
7	0.6	3,4
2	0.4	—
5	0.3	2
8	0.3	5
6	0.2	3
1	0.15	—
4	0.1	1,2

Table 9.3   Work Elements in Descending Order of Processing Times.

To determine the work elements to be assigned to the second station, we again start at the top of the list in Table 9.3 with the previously chosen work elements removed. Going down the list we see that the first appropriate element is Element 3, and the next is Element 6. Selecting additional elements exceeds 1.0 min, and thus the work elements to be processed at the second station are Elements 3 and 6. This leaves Elements 7 and 8 for the third, and last, station. Hence, the total number N of stations is equal to 3.

In this solution, the bottleneck station (not including breakdowns) is the first station since it has the longest processing time, which is 0.95 min. The processing time of both the second and third stations is equal to 0.9 min. The cycle time is therefore equal to 0.95 min, and from (9.35) the balance delay d is

$$d = \frac{(3)(0.95) - 2.75}{(3)(0.95)} = 0.035$$

Since the value of d is very small, this solution for the arrangement of the processing elements yields a line configuration that will have only a small amount of idle time. Since the processing time of the first station is 0.95 min and the processing time of both the second and third stations is 0.9 min, in every cycle the second and third stations are both idle for 0.05 min.

Problems

9.1 A product is manufactured by processing a part at six different machines. The parts are processed in batches of 25. The operation time per part per machine is 6 min, and the setup time and nonvalue-adding operation time per machine per batch is 2 hr, and 4 hr, respectively. There are 10 machines in the manufacturing plant that operates a total of 70 hr per week.

(a) Compute the MLT for producing the first batch.

(b) Compute the maximum possible production rate in parts processed per machine per hour.

(c) Compute the capacity in parts processed per week.

(d) Compute the capacity in products produced per week.

(e) Compute the WIP.

9.2 The production of a product is carried out using a flow line consisting of six machining operations with the processing times in minutes equal to 0.4, 0.5, 0.8, 1.2, 1.5, and 1.6. The batch size is equal to one and the setup times are neglected in all of the parts given below. Suppose that there are six workstations in the production facility, a part must visit each workstation, and the transfer time from station-to-station is ten seconds. Neglecting breakdowns, determine:

(a) The production rate given in terms of the number of products per hour

(b) The capacity equal to the number of products per 48-hr week

(c) The WIP

9.3   For the production facility in Problem 9.2, now assume that two of the workstations break down with probability 0.01, two break down with probability 0.02, and other two break down with probability 0.03. Also assume that there are no buffers between the stations and that the time to fix a breakdown is 10 minutes.

(a) Compute:

(i)   The production rate given in terms of the number of products per hour

(ii)   The capacity equal to the number of products per 48-hr week

(iii) The WIP

(iv)   The line efficiency E_o

(b) Suppose that one buffer with infinite capacity is available to be placed after any one of the machines in the production facility. Put the buffer in the location that will yield the maximum efficiency. Specify the location of the buffer and compute the new line efficiency.

(c) Repeat part b assuming that two infinite-capacity buffers are available.

9.4   For the production facility in Problem 9.2, suppose that there are only three workstations and that each of the stations can perform all six machining operations. Assume that each station processes parts one at a time, the transfer time between stations is still 10 sec, and that stations do not break down. Determine:

(a) The production rate given in terms of the number of products per hour

(b) The capacity equal to the number of products per 48-hr week

(c) The WIP

9.5   The production of a product is carried out using a flow line consisting of four machining operations with the processing times in minutes equal to 1.5, 3.5, 5.0, and 2.0. The batch size is equal to one and the setup times and transfer times are neglected in all of the parts given below. Suppose that there are no breakdowns and parts are put into the system one right after the other with no parts in the system prior to the application of Part 1.

(a) Compute the MLT for Part 1, Part 2, and Part 3.

(b) Compute the production rate given in terms of the number of finished products per hour.

(c) Compute the WIP and the capacity equal to the number of products per 48-hr week.

9.6   For the production facility in Problem 9.5, suppose that the machines can break down with probability 0.01 for the machine with processing time 5.0, 0.02 for the machine with processing time 3.5, and 0.03 for the machines with processing times 1.5 and 2.0. Also assume that the time required to fix a breakdown is 5 min.

(a) Assuming there are no buffers, compute the production rate in finished products per hour and the line efficiency.

(b) Assuming there are infinite-capacity buffers before each machine, compute the production rate in finished products per hour and the line efficiency

(c) Assuming that the processing order is fixed by the sequence 1.5, 3.5, 5.0, and 2.0 and that one buffer with infinite capacity is available to be placed anywhere along the line, put the buffer at the location that will yield the greatest line efficiency. Specify the buffer location and compute the new line efficiency.

(d) Rather than having separate machines that perform each of the four operations, the company can purchase workstations that are capable of performing all four operations. Assuming that a workstation processes one part at a time, stations do not break down, and the processing order is arbitrary, determine the maximum possible production rate in products per hour in these situations:

(i) Only one workstation is used.

(ii) Two workstations are used.

(iii) Three workstations are used.

9.7 A product is manufactured by processing a part through six machines. The parts are processed in batches of 25. The operation time per machine is 6 min and the setup time and nonvalue-adding operation time per machine per batch is 2 hr, and 4 hr, respectively. The manufacturing plant operates a total of 70 hr per week.

(a) Compute the MLT for manufacturing 25 products.

(b) Compute the MLT for manufacturing 50 products.

(c) Compute the MLT for manufacturing 75 products.

(d) Compute the MLT for manufacturing 100 products.

(e) Compute the capacity in parts produced per week.

(f) Compute the capacity in products produced per week.

(g) Compute the WIP assuming the batches are processed one after another.

(h) Determine the maximum possible production rate (products per hour).

9.8 A manufacturing transfer line contains two machines (operating under the FIFO policy) with infinite-capacity buffers before each machine. Parts are processed one-at-a-time on each machine. The input to the line is $A(k)$ = input time of Part k, and the output of the line is $C(k)$ = output time of Part k. The service time of the first machine is 5 min and the service time of the second machine is 10 min. In all of the following parts, assume that there are no parts in the system at time $t = 0$, and $C_1(1)$ = time of first output of the first machine = 5 min, $C(1)$ = time of first system output is = 15 min.

(a) When $A(k) = 5(k-1)$, compute $C(k)$ for $k = 2, 3, 4, 5$.

(b) When $A(k) = 10(k-1)$, compute $C(k)$ for $k = 2, 3, 4, 5$.

(c)   When $A(k) = 15(k-1)$, compute $C(k)$ for $k = 2,3,4,5$.

(d)   Compare your results in parts (a), (b), and (c). What do you conclude?

(e)   Assuming that $A(1) = 0$ and $A(k) = C(k-1)$ for k = 2,3,4,..., compute $C(k)$ for $k = 2, 3, 4, 5$.

(f)   Compare your results in part (e) with the results in parts (a), (b), and (c). What do you conclude?

(g)   For Parts (a), (b), (c), and (e), how many parts are in the system at time $C(5) - 1$ min?

9.9   The manufacturing of a product requires that a part be processed using three machines M_1, M_2, and M_3 which operate under the FIFO policy. Parts are put into the manufacturing line one at a time, and are processed by the machines in the sequence M_1, M_2, and M_3. The processing times in minutes of the machines are $M_1 = 5$, $M_2 = 12$, and $M_3 = 15$. The processing times include transfer times. Setup times are neglected. It is assumed that each machine operates on a part as soon as it arrives (at the machine) if the machine is not already busy processing another part. The kth part is applied to the manufacturing line at time $A(k)$, and the kth completion time after processing by all three machines is equal to $C(k)$. Both $A(k)$ and $C(k)$ are in minutes. It is assumed that at time $t = 0$ there are no parts in the manufacturing line.

(a)   Assuming that there are infinite-capacity buffers before each machine and $A(k) = 5(k-1)$, compute $C(k) - A(k)$ for $k = 1, 2, 3$.

(b)   Assuming that there are infinite-capacity buffers before each machine and $A(k) = 5(k-1)$, determine the number of parts in each buffer at time $C(3) + 1$ min.

(c)   Assuming that the buffers before the machines can hold only one part, what is the fastest possible MLT for part 1, part 2, and part 3. Express your answer in minutes.

9.10   The production of a product is carried out using a flow line consisting of five operations with the processing times in minutes equal to 0.2, 0.4, 0.6, 0.1, and 0.5. The batch size is equal to one and the setup times and transfer times are neglected The operations have the following precedence constraints:

Operation number	Processing time	Must be preceded by operation
1	0.2	—
2	0.4	3
3	0.6	1
4	0.1	3
5	0.5	4

(a) Using the largest candidate rule to achieve $T_c = 0.9$ min, determine the smallest number of workstations needed, and specify the operations to be performed on each station.

(b) Determine the "goodness" of your solution in part (a).

(c) For your solution in part (a), determine the idle time in a given cycle at each of the workstations.

Chapter 10

Production Control

An important component of manufacturing is production control, which involves the manner in which the production facility is operated in order to meet desired performance requirements such as satisfying customer demand in a timely manner. Production control is a higher level of control in comparison to process control that was considered in Chapters 4 through 8, and thus production control is quite different from process control, although as will be seen in this chapter there are some similarities.

To define what production control is, consider the flow line shown in Figure 10.1. In this figure, $A(k)$ denotes the time when the kth part is taken from the raw materials inventory and put into the system, and $C(k)$ is the time when the line has completed processing of the kth part and the finished part (i.e., the product) is put into the product inventory. The values of $A(k)$ are often referred to as the *release times* since they are the times when the raw materials are released into the system. If we assume that the stations or machines within the line operate on parts as soon as the parts are available whenever the stations are not already busy, then production control for the line reduces to the problem of determining the release times $A(k)$ so as to achieve a desired performance. Hence, as in process control, the production control problem involves the selection of appropriate inputs, which in this case are the release times $A(k)$. However, in contrast to process control, in production control the inputs that can be applied are very restricted. In particular, as seen from the analysis in the previous chapter, it is not possible to select any desired values for $A(k)$ since a part can be put into the line only if the line has "room" to accept it. This constraint greatly complicates the production control problem.

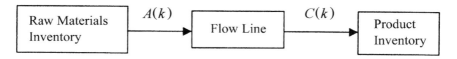

Figure 10.1   Flow line with raw materials inventory and finished
products inventory.

Production Control Schemes

The simplest type of production control scheme is to operate the line at full capacity or not to operate the line. This is the "full-on" or "full-off" mode of operation. In the full-on mode, a part is put into the system whenever the line can
accept another part. If there is a buffer before the first station in the line, this
means that the first buffer has space for another part.

In the full-on or full-off mode of operation, the production facility is usually operated at full capacity during the normal working hours of the plant, and
is off (shut down) otherwise. The primary problems with this type of operation
are:

1. It is not responsive to varying customer demands.
2. It may result in large WIP and product inventory.
3. The production rate is often highly variable due to equipment
    breakdowns, blocking, starvation, and time-varying parameters such
    as varying processing times and varying material transport times.

The large variation of the production rate that can occur in full-on operation is a
major problem because it is difficult to predict the production output when there
is a substantial amount of variability. Regardless of the production control
scheme that is used, variability is always present to some extent. It can be seen
in the charts that manufacturers generate to track the production output per day
or per week. Later we will see that there are production control schemes that
tend to smooth out variations.

In addition to the variability in the production rate, operation in the full-on
mode can produce a large WIP as a result of "stuffing" the line with parts. To
reduce the congestion that can occur in full-on operation, the rate at which parts
are put into the line can be set at some fractional value of the maximum possible
production rate $(R_p)_{max}$. As discussed in the previous chapter, $(R_p)_{max}$ can be computed by taking the inverse of the average processing time at the bottleneck sta-

tion (with the effect of breakdowns included). The input rate can then be set to $\mu(R_p)_{max}$, where $0 < \mu < 1$; that is,

$$\frac{1}{A(k) - A(k-1)} = \mu(R_p)_{max} \qquad (10.1)$$

where $A(k)$ is the time when the kth part is put into the line. With the input rate given by (10.1), the average value of the output rate (the production rate) will be equal to $\mu(R_p)_{max}$ if the system is *stable* (i.e., parts in the buffers are not growing without bound). The parameter μ can be determined by operating the line with different values for μ. In general, the closer μ is to 1, the greater the congestion in the line, which can result in degraded performance in terms of meeting a desired production rate.

Ideally, production should be based on customer demand. One approach to achieving this is to use *material requirements planning* (MRP). In MRP, the due dates of finished products are used to schedule production; in other words, the due dates are used to determine the release times $A(k)$. This is accomplished by computing the manufacturing lead time (MLT), which was defined in the previous chapter. The computation of MLT for a flow line can be carried out as follows.

Suppose that the average desired production rate of the line is $R_p = \mu(R_p)_{max}$ products per hour. Then after the first finished part (i.e., product) leaves the line, a product will exit the line every $1/R_p$ hours on the average. Thus, the average value of the MLT required to manufacture P products is given by

$$\text{MLT} = \frac{P-1}{R_p} + T_1 \qquad (10.2)$$

where T_1 is the time required to produce the first finished part. The quantity T_1 includes the line setup time and the time for the processing and movement of a part as it goes through the entire line.

Then given the due date for the quantity of P products, (10.2) can be used to determine when production should begin. When production begins, the input rate is set equal to R_p; that is,

$$\frac{1}{A(k) - A(k-1)} = R_p \qquad (10.3)$$

Production then continues over the time interval specified by the MLT with the input rate given by (10.3). Although this is a rather simplistic view of how MRP works, one should have some idea as to how the scheduling of release times is accomplished based on customer demand.

The full-on mode and the scheduling of release times are examples of a type of production control system called a *push system*. In a push system the release times are set based on some criterion such as customer demand and are independent of the status of the system. Thus, a push system is similar to an open-loop process control system where the control input does not depend on the process variables or the process output.

In a push system the production rate is controlled by setting the input rate. However, since the release times are independent of the system status, the performance of push systems can be very sensitive to variations in the system such as changing processing times. To achieve robustness to variability in the system, one can consider another type of production control system referred to as a *pull system*. In a pull system releases are authorized based on the status of the line, and thus a pull system is similar to a closed-loop process control system where the control input depends on the process variables or the process output. However, in a pull system it is not possible to control the production rate directly. We study two types of pull systems next and then we consider a system that is a combination of push and pull.

Pull Systems

An example of a pull system is CONWIP, which stands for "constant work-in-process." As the name implies, in a CONWIP system the WIP in the system is kept constant at all times. In a CONWIP system, Part k is put into the system when the line has finished processing Part $k - W$, where W is the desired WIP level. Hence, in terms of the notation in Figure 10.1, the control is given by

$$A(k) = C(k - W) \qquad (10.4)$$

Using (10.4) we have that

$$A(k) - A(k - 1) = C(k - W) - C(k - W - 1)$$

and, thus, the input rate in a CONWIP system is given by

$$\frac{1}{A(k) - A(k-1)} = \frac{1}{C(k-W) - C(k-W-1)} \qquad (10.5)$$

From (10.5) we see that the input rate is a "delayed version" of the output rate. If the output rate (which is the same as the production rate) is equal to a constant R_p, then in a CONWIP system the input rate is equal to R_p.

It is important to note that by using (10.4), we can control the WIP but we cannot control the production rate. If (10.4) is used, the production rate R_p will be whatever it must be to maintain a constant WIP. If R_p is not varying a great deal, it can be computed using Little's law, which states that

$$\text{WIP} = R_p[C(k) - A(k)] \qquad (10.6)$$

Inserting (10.4) into (10.6) and solving for R_p yields

$$R_p = \frac{\text{WIP}}{C(k) - C(k-W)} \qquad (10.7)$$

Since WIP is constant in a CONWIP system, from (10.7) we see that any variability in R_p is a result of variability in $C(k) - C(k - W)$. Obviously, variability in WIP would add to the variability of R_p, and thus CONWIP does tend to smooth out variations in the production rate in comparison to pull systems where WIP is not controlled.

To employ CONWIP, it is necessary to specify the desired value W of the WIP. If the line contains N stations, it would seem to be reasonable to set $W = N$ so that every station will be busy all the time. But if the processing times at the stations are not balanced (not approximately the same), not all of the WIP will be under processing at the stations – part of the WIP will be waiting for processing. Hence, it may seem reasonable to take the value W of the WIP to be less than N. A lower bound on the choice of the WIP is the critical WIP, which was first defined in Chapter 9 and is given by

$$\text{CWIP} = \frac{\sum_{i=1}^{N} T_{pi}}{max(T_{p1}, T_{p2}, ..., T_{pN})}$$

where T_{pi} is the average processing time at the ith station including the effect of breakdowns. Usually the value of the WIP is chosen to be strictly greater than the value of the critical WIP since the computation of CWIP is based on ideal conditions that are unlikely to exist in practice (see Chapter 9).

In general, we would like to have parts in the buffers before the stations to smooth out variations due to breakdowns, and thus we may want to choose the value of the WIP to be greater (perhaps much greater) than N. Hence, the "appropriate value" of the WIP could be greater than or less than N depending on the desired performance of the line. The best value for the WIP may be determined by trial and error while operating the line.

It turns out that the analysis of a CONWIP system is fairly complicated due to the constraint that the WIP be constant. We now illustrate this by considering a line with two machines.

Two-Machine Case

Suppose that the line consists of two machines with buffers as illustrated in Figure 10.2. The buffers are assumed to have some finite capacity. As indicated in the figure, $C_1(k)$ is the kth completion time of M_1; that is, $C_1(k)$ is the time when the kth part has completed processing at M_1. To derive an expression for $C_1(k)$, we first determine the kth start time for M_1. The difficulty here is that the kth start time of M_1 cannot occur until there is room in the second buffer to accept the output of M_1 when the kth part is done. If we force the WIP in the second buffer and M_2 to be equal to W_2, then the kth start time of M_1 cannot occur until $C(k - W_2)$, which is the time when Part No. $k - W_2$ leaves the second machine (and the line). Thus the kth start time of M_1 is equal to

$$max[C_1(k-1), A(k), C(k-W_2)]$$

and

$$C_1(k) = max[C_1(k-1), A(k), C(k-W_2)] + T_{p1} \qquad (10.8)$$

where T_{p1} is the processing time at the first machine. Then to keep the WIP in the system equal to W where $W > W_2$, we must set $A(k) = C(k - W)$, and therefore (10.8) becomes

$$C_1(k) = max[C_1(k-1), C(k-W), C(k-W_2)] + T_{p1}$$

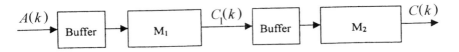

Figure 10.2 Line with two machines.

But since $W > W_2$, $C(k - W) < C(k - W_2)$; thus,

$$C_1(k) = max[C_1(k-1), C(k-W_2)] + T_{p1} \qquad (10.9)$$

The kth completion time $C(k)$ of M_2 is given by

$$C(k) = max[C(k-1), C_1(k-1)] + T_{p2} \qquad (10.10)$$

where T_{p2} is the processing time at M_2. Equations (10.9) and (10.10) describe the two-machine line under CONWIP control with the additional constraint that the WIP in the second buffer and machine is equal to W_2. Since the WIP is set to the value W, the WIP in the first buffer and machine is equal to $W - W_2$.

The production rate of the two-machine system can be determined by solving (10.9) and (10.10). To illustrate this, suppose that $T_{p1} = 1$ min, $T_{p2} = 2$ min, and we set $W = 4$, $W_2 = 2$. We also assume that $C_1(1) = 1$ and $C(1) = 3$. Since there is no $C(0)$, setting $k = 2$ in (10.9) and (10.10) gives

$$C_1(2) = 1 + 1 = 2$$

$$C(2) = max(3,2) + 2 = 5$$

Setting $k = 3, 4, 5, 6$ in (10.9) and (10.10) yields

$$C_1(3) = max(2,3) + 1 = 4$$

$$C(3) = max(5,2) + 2 = 7$$

$$C_1(4) = max(4,5) + 1 = 6$$

$$C(4) = max(7,4) + 2 = 9$$

$$C_1(5) = max(6,7) + 1 = 8$$

$$C(5) = max(9,6) + 2 = 11$$

$$C_1(6) = max(8,9) + 1 = 10$$

$$C(6) = max(11,8) + 2 = 13$$

Hence, for $k = 1, 2, 3, 4, 5, 6$, $C_1(k) = 1, 2, 4, 6, 8, 10$ and $C(k) = 3, 5, 7, 9,$ 11, 13. Note that the first machine processes the first two parts at the rate of one per minute, but after the second part, the processing rate is one part every 2 min. The reason for the change in processing rate is due to the first machine being blocked when the WIP is equal to 2 in the second buffer and machine. That is, the first machine cannot start processing a new part when the WIP is equal to 2 in the second buffer and machine. Note also that the production rate of the two-machine line is equal to the production rate of the second machine, which is the bottleneck machine.

Kanban

Another type of pull system is *kanban*, which was developed in Japan by Toyota. In a kanban system, each stage (consisting of a station and buffer) of the production line has a set of cards, which are called *kanbans* in Japanese. When a part arrives at the ith stage, a card is attached to it until it leaves that stage for the next one. If all kanbans (cards) at the ith stage are attached to parts in that stage, no new parts can be admitted to the ith stage. When a part in the ith stage is admitted to the $(i+1)^{th}$ stage, the kanban attached to the part is removed and is then available to be attached to a new part entering the ith stage.

From the description of kanban, it is clear that the WIP level is controlled at each stage, and thus kanban is a type of local inventory control. Since the WIP is controlled at each stage, the overall WIP in the system is controlled as in CONWIP. However, kanban is a more constrained type of production control than CONWIP since in the former case the WIP at each of the stages is also controlled.

The movement of parts through a kanban system can be modeled as follows. First, suppose that the processing time of the ith stage is T_{pi}, and there are q_i cards at the ith stage. Let $C_i(k)$ denote the time when the ith stage is finished processing Part k. The ith stage can admit the kth part only after it has finished processing Part $k - q_i$, the $(I - 1)^{th}$ stage has finished processing the kth part, and the $(i+1)^{th}$ stage has finished processing Part $k - q_{i+1}$. Hence, the kth part can be admitted to i^{th} stage at time

$$max[C_i(k - q_i), C_{I-1}(k), C_{i+1}(k - q_{i+1})]$$

and the ith stage will be finished processing the kth part at time

$$C_i(k) = max[C_i(k - q_i), C_{I-1}(k), C_{i+1}(k - q_{i+1})] + T_{pi} \qquad (10.11)$$

The flow of parts through the kanban system can be studied by evaluating (10.11) for $i = 1, 2, ..., N$ where N is the number of stages in the system.

A Push-And-Pull System

As noted previously, pull systems do not control WIP, and pull systems such as CONWIP and kanban do not control the production rate. Ideally, we would like to control both WIP and the production rate based on customer demand. This can be accomplished by controlling the production rate based on inventory levels. This type of production control system, which is described later, is both push and pull. We begin by considering *inventory control.*

Suppose that the output of a production line goes into an inventory and that the number of products in the inventory at time $t = nT$ is equal to $X(nT)$. Here $n = 0, 1, 2, ...$ is an integer index and T is a fixed interval of time (e.g., $T = 1$ hr or 1 day). We let $\mu(nT)$ denote the number of products added to the inventory in the time interval from nT to $nT+T$, and we let $\lambda(nT)$ denote the number of products taken from the inventory for shipment to customers in the time interval from nT to $nT+T$. Note that the dimensions of both $\mu(nT)$ and $\lambda(nT)$ are products per unit of time T. Clearly, the rate $\mu(nT)$ at which products are added to the inventory is equal to the production rate of the line, and the rate at which products are removed from the inventory depends on the customer demand. The range of possible values for $\mu(nT)$ is

$$0 \leq \mu(nT) \leq (R_p)_{max}$$

where $(R_p)_{max}$ is the maximum production rate of the line in products per unit of time T. So the variable $\mu(nT)$ is constrained to lie in this range.

The number $X(nT+T)$ of products in the inventory at time $nT+T$ is given by

$$X(nT + T) = X(nT) + \mu(nT) - \lambda(nT) \qquad (10.12)$$

The control objective is to vary the production rate $\mu(nT)$ of the line so that the number $X(nT)$ of products in the inventory is kept at a desired level Z, called the *hedging point*. One possible "control law" is as follows:

$$\text{if } X(nT) < Z, \quad \text{then set } \mu(nT) = (R_p)_{max}$$

$$\text{if } X(nT) = Z, \quad \text{then set } \mu(nT) = \lambda(nT - T)$$

$$\text{if } X(nT) > Z, \quad \text{then set } \mu(nT) = 0$$

This control is rather simplistic, but can yield reasonable performance. There are more complicated controls that are based on results from "constrained optimal control," but this is beyond the scope of this book.

WIP Control

The inventory control scheme can be extended to control the WIP at the stages in the production line: Let $X_i(nT)$ denote the number of parts in the ith stage at time nT and let $\mu_i(nT)$ denote the number of parts processed by the ith stage in the time interval from nT to $nT+T$. The quantity $\mu_i(nT)$ is the production rate of the ith stage in parts per time interval T. Then we can represent the ith stage in terms of the relationship

$$X_i(nT + T) = X_i(nT) + \mu_{i-1}(nT) - \mu_i(nT + T)$$

Beginning with the last stage ($i = N$), the control objective is to vary the production rate $\mu_i(nT)$ of the ith stage so that the number $X_i(nT)$ of parts in the ith stage is kept at a desired level Z_i, which is the desired level of WIP for that stage. A simple control strategy to achieve this is as follows:

$$\text{if } X_i(nT) < Z_i, \quad \text{then set } \mu_i(nT) = \text{maximum production rate for Stage } i$$

$$\text{if } X_i(nT) = Z_i, \quad \text{then set } \mu_i(nT) = \mu_{i-1}(nT - T)$$

$$\text{if } X_i(nT) > Z_i, \quad \text{then set } \mu_i(nT) = 0$$

This control is not optimal with respect to variations. As noted, before, more complicated controls exist, but we do not pursue this.

Problems

10.1   A production flow line consists of seven stations with the average processing times (including breakdowns and transfer times) at the stations equal to 4.5, 6.1, 7.2, 4.1, 7.6, 5.7, and 6.8 min. It takes 2 hr to set the line up for manufacturing a product. If a customer order is received for a quantity of 100 of the product, what is the latest time at which production must begin in order to fill the order by the due date? Assume that the production facility can be operated 24 hours a day.

10.2   A manufacturing transfer line contains two machines (operating under the FIFO policy) with infinite-capacity buffers before each machine. Parts are processed one-at-a-time on each machine. The input to the line is $A(k)$ = input time of Part k, and the output of the line is $C(k)$ = time of kth system output. The service time of the first machine is 5 min and the service time of the second machine is 10 min. In all of the following parts, assume that there are no parts in the system at time $t = 0$, and $C_1(1)$ = time of first output of Machine 1 = 5 min, and $C(1)$ = time of first system output is = 15 min.

(1) Determine $C(k)$ for $k = 1, 2, 3, 4, 5, 6$ when the system is operated using CONWIP with WIP = 1.

(2) Repeat part (a) with WIP = 2.

(3) Repeat part (a) with WIP = 3.

(4) Determine the steady-state production rate under CONWIP operation with

(i)   WIP = 1

(ii) WIP = 2

(iii) WIP = 3

(5) Based on your results in part (d), what do you conclude?

10.3 Repeat Problem 10.2 with the two machines interchanged so that the machine with the longer processing time is first.

10.4 Again consider the two-machine setup in Problem 10.2.

(a) Determine $C(k)$ for $k = 1, 2, 3, 4, 5, 6$ when the system is operated using kanban with one card for Stage 1 and one card for Stage 2.

(b) Repeat part (a) with two cards for each of the stages.

(c) Repeat part (a) with three cards for each of the stages.

(d) Determine the steady-state production rate under kanban operation with

(i)   One card at each stage

(ii) Two cards at each stage

(iii) Three cards at each stage

(e) Based on your results in part (d), what do you conclude?

10.5 Repeat Problem 10.4 with the two machines interchanged so that the machine with the longer processing time is first.

10.6 For the two-machine setup in Problem 10.2, suppose that the output of the line is placed into a product inventory. When the input rate is one part every 15 min and $\lambda(nT) = 1, 3, 2$ when $n = 0, 1, 2$, compute the number $X(nT)$ of products in the product inventory at the times T, $2T$, and $3T$, where $T = 1$ hr. Assume that $X(0) = 0$.

10.7 For the two-machine setup in Problem 10.2, use the control law given in this Chapter to adjust the input rate so that the hedging point Z is equal to 5 assuming that $X(0) = 0$ and $\lambda(nT) = 1, 3, 2$ when $n = 0, 1, 2$ and $T = 1$ hour. Express your answer by giving the values of $A(k)$ for k ranging from 1 to 12.

Chapter 11

Equipment Interfacing and Communications

As first mentioned in Chapter 1, an important part of the overall control problem is the interfacing and the communications between controllers, sensors, devices, and graphics terminals in the system being controlled. In fact, the performance and flexibility of a control system very much depend on how the interfacing and communications are carried out. In this chapter we give a brief introduction to interfacing and communications between the components in a system. We begin by considering the interfacing between equipment and a host.

Equipment Interfacing

The basic components of a connection to an equipment item such as a sensor or a device are shown in Figure 11.1. The block in the figure titled "Host" could be a controller such as a PLC, or it could be a PC, a computer workstation, or a graphics terminal. The block titled "Application" represents a software module that provides instructions to the host for carrying out some sequence of tasks involving data processing, control signal generation, etc. Since the host serves (i.e., implements) the instructions provided by the application, the host is sometimes referred to as the *server*, and the application is referred to as the *client*. As a result, the block diagram in Figure 11.1 is often referred to as a *client/server architecture*.

205

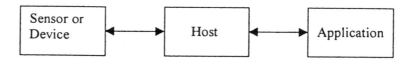

Figure 11.1 Connection to an equipment item.

The double arrow in Figure 11.1 between the equipment item and the host indicates that data can flow from the equipment to the host or from the host to the equipment. In some cases, the data flow is only from the equipment to the host. This is often the case when the device is a sensor that simply sends measurement data to the host. If the data flow is bidirectional, the device is sometimes said to be *intelligent* or *smart*. There exist smart sensors that accept commands from the host. An example is a sensor that sends data to the host only when requested to do so by the host, or by the application that is installed on the host.

In Figure 11.1 the double arrow between the host and the application indicates that information can go back and forth between the host and the application. The flow of information from the application to the host consists of commands and instructions to the host. In the other direction, information in the host may be used by the application to generate specific instructions to the host. For instance, based on the status of the equipment, the application may request that a specific set of data be acquired.

A key issue in the interfacing of equipment to a host is the manner in which the data that flow between the equipment and the host are formatted and addressed. It has been common practice for different vendors of equipment to use their own proprietary methods for handling data, and as a consequence, a special type of interface is required to connect their equipment to a host such as a PC or workstation. This has resulted in a lack of interchangeability or interoperability of equipment from different vendors.

At the time of writing of this book, efforts are under way to develop a set of standards that describes how data are to be formatted and addressed in the interface between equipment, host, and applications. One such standard is OLE for process control (OPC), which is based on object linking and embedding (OLE) and the component object model (COM) developed by Microsoft. OPC provides a common interface for interconnecting applications to various process control devices. Version 1.0 of the OPC standard was released in August 1996 and is under continued development by the OPC Foundation. Technical committees in the OPC Foundation are working on various extensions of the OPC standard including the following:

- Allowing applications to exchange blocks of historical data
- Defining how alarms are to be handled
- Defining how events are to be communicated
- Defining a standard for naming data in control devices

The status of the extensions to the basic OPC standard can be checked by contacting the OPC Foundation web site at www.opcfoundation.org.

Any device and any application that is OPC compliant (i.e., is based on the OPC standard) can be directly connected together through an OPC server (or host) without having to write code or develop a special driver. As a result, the standard provides for a "plug-and-play" capability in that devices or application software modules from different venders can be interchanged without having to modify either software or hardware.

GEM Standard

In electronics manufacturing a standard called the *generic equipment model* (GEM) has been established to facilitate the interfacing of equipment with a host. The GEM standard is in use in both semiconductor wafer manufacturing and in printed circuit board electronics assembly. The standard is based on a SECS-II communication environment, where SECS stands for "semiconductor equipment communications standard." The SECS standard defines the messages that are passed between equipment and a host. All messages are labeled. Examples of SECS-II messages include:

S1,F1   Are you there request
S2,F13   Equipment constant request
S5,F1   Alarm report send
S6,F1   Trace data send
S7,F5   Process program request

The GEM standard defines the general behavior of equipment as seen by the host. Equipment behavior is specified in terms of state diagrams that describe various aspects of operation including communications between equipment and the host and the degrees of control that the host can exert over the equipment. The state diagrams are specified in terms of a hierarchy of states; that is, states may have substates, and the substates may have substates. This is in contrast to the state diagrams considered in Chapters 7 and 8 where substates were not used. For example, a portion of the state diagram for the control part of GEM is given in Figure 11.2. As seen from the figure, there are two states: Off-Line and On-Line, and each state has "OR" substates. When a state with OR substates is active, the process is always in one of the substates. Hence, when the Off-Line state is active, the process is in the Equipment Off-Line, Host Off-Line, or Attempt On-Line substate, and when the On-Line state is active, the process is in either the Local or Remote substate. When the On-Line/Remote substate is active, the host may operate the equipment to the maximum extent possible. When the On-Line/Local substate is active, operation of the equipment is accomplished via an operator and the host has a very limited capability in controlling the equipment. In the Off-Line/Equipment Off-Line substate, the host has no control over the equipment. In the Off-Line/Host Off-Line substate, the intent is that the equipment be On-Line but the host has not agreed to the change to the On-Line state, and in the Off-Line/Attempt On-Line substate, the equipment is in the process of going On-Line.

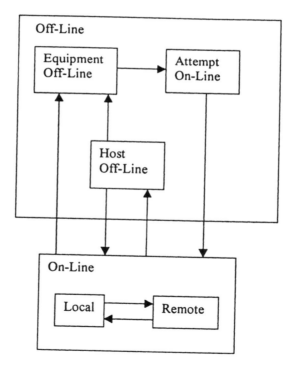

Figure 11.2 State diagram for control part of GEM.

The GEM standard provides the capability for data to be sent to the host at specified points during equipment operation. This is accomplished through the use of collection events, which are events occurring on the equipment that are of importance to the host. Each event is associated with one or more event reports that contain data relating to the event. The data contained in an event report may consist of status variables, equipment constant values, or data values. The host can also request trace data collection which is based on the periodic sampling of variables associated with events. This capability allows the host to monitor the status of the equipment for diagnostic purposes.

Communications

For many years the communication between devices and controllers was based on a 4-20 milliampere (mA) standard for analog signals. Devices were "hardwired" to controllers in order to allow communication between the two. However, as the complexity of applications increased, the number of devices that had to be connected to a single controller became so large that it was not feasible to wire each device to the controller. Even for PLCs that have a large number of input–output connections, it is often the case that the devices are distributed in the field away from the PLC so that connecting each device to the PLC would result in an excessive amount of wiring.

The solution to this wiring problem was to connect the devices and the controller to a network with digital links consisting of twisted-pair wire, coax cable, or optical fiber cable. The three basic topologies (layouts) of a network are bus, ring, and star, which are illustrated in Figure 11.3. In the bus and ring topologies, the components are connected to the network through medium attachment units (MAUs), while in the star topology, the components are connected to a central hub consisting of a repeater or switch. Originally, the bus topology was dominant, but today the star topology is much more common. Part of the reason for this is the enhanced reliability that can be achieved with the star topology. In particular, in the bus and ring frameworks the entire network can go down if any one of the MAUs fails; whereas in the star framework the connection of a component to the central hub can fail without affecting the other components connected to the hub.

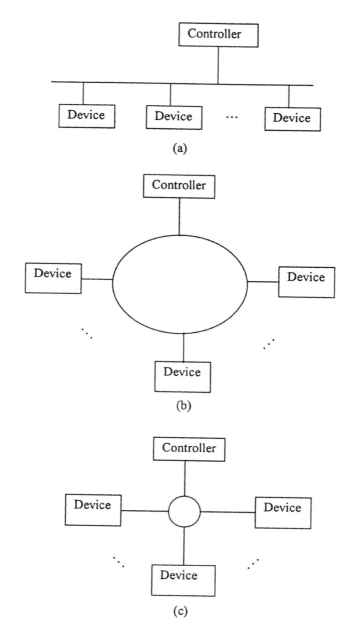

Figure 11.3 Three network topologies: (a) bus, (b) ring, and (c) star.

Network Protocols

The operation of communication networks is based on a set of formalized procedures called *protocols* that describe how data or information is exchanged among the devices connected to the network. To simplify the specification of the protocols, the operation of communication networks is partitioned into layers that describe the various aspects of functionality relating to the exchange of data and information among the devices on the network. A standard for this layered approach is the Open Systems Interconnection (OSI) reference model that was developed by the International Standards Organization. The OSI model consists of the seven layers shown in Figure 11.4. A brief description of each layer follows:

1. *Physical:* Deals with bit transfer and signaling, defines the type of transmission medium between stations (devices) on the network and the rules by which bits are sent from one station to another.
2. *Link:* Makes the physical link reliable, provides the means to activate, maintain, and deactivate links, and specifies how the network is to be accessed by the stations on the network.
3. *Network:* Accepts messages from stations, segments them into packets, and makes sure that packets get to their destination.
4. *Transport:* Provides a reliable mechanism for the exchange of data between stations, insures that data is delivered error free, in sequence, and with no missing data or duplications.
5. *Session:* Provides dialogue management between the applications running on the stations.
6. *Presentation:* Resolves differences in format and data representation between applications running on stations.
7. *Application:* Provides the interface with applications running on stations.

7. Application
6. Presentation
5. Session
4. Transport
3. Network
2. Link
1. Physical

Figure 11.4   The seven layers of the OSI model.

Most communication networks do not use the specific protocols of the OSI model, but the seven-layer format is still very useful when comparing the protocol structure of networks. For example, sensor and device networks employed in control have only three layers that correspond to Layers 1, 2, and 7 of the OSI model. Examples of sensor and device networks include BitBus, Seriplex, ArcNet, DeviceNet, and CAN (controller area network). The primary feature that distinguishes these networks from each other is the manner in which devices on the network can access the network. This is called *medium access control* (MAC), which is resolved in Layer 2 of the protocol structure. There are four basic types of MAC:

- Master–slave
- Consultation
- Contention
- Master clock (synchronous)

In the master–slave protocol, one device is defined to be the master that controls when the other devices can transmit messages. The MAC for DeviceNet is based on the master–slave type. An example of a consultation-based MAC is token passing where a special code, called the *token*, is passed along the network at high speeds in a predetermined sequence. A station can transmit a message only when it possesses the token. If a station has no message to send, the token passes to the next station in the sequence. If a station has a message to send, it must wait until it has possession of the token before it can transmit. ArcNet uses the token passing protocol.

In a contention-based MAC, the stations must compete to gain access to the network. An example of this type of MAC is *carrier sense multiple access with collision avoidance* (CSMA/CR), which is utilized by CAN. In the master clock mode, access to the network is based on a predetermined timing sequence. Seriplex is based on a master clock MAC.

Process Networks/Fieldbuses/Ethernet

Networks that interconnect process controllers, PLCs, analyzers, graphics terminals, etc., are called *process networks, control networks,* or *fieldbuses.* This type of network, which we will call a fieldbus, is one level higher than a sensor or device network. Here the term "level" refers to the degree of sophistication (i.e., intelligence) of the devices connected to the network at that level: The higher the level, the more sophisticated the devices are.

Examples of fieldbuses are Profibus, Lon Works, ControlNet, and Fieldbus. Also, there is a growing emphasis on the use of *Ethernet* as a fieldbus. Ethernet was developed more than 20 years ago for use as a *local-area network* (LAN) for interconnecting computers. Its protocol structure has four layers that correspond to Layers 1, 2, 3, and 4 of the OSI model. The MAC (Layer 2) for Ethernet is based on *carrier sense multiple access with collision detection* (CSMA/CD). In this protocol, which is defined in IEEE Standard 803.2, stations listen for activity on the network. If the network is busy, a station will hold off transmitting until the network is idle. Otherwise, a station will send its message. If two or more stations transmit at the same time, the messages will collide, resulting in the termination of all the transmissions. In this case, the stations wait a random length of time before attempting to retransmit.

For Layer 3, Ethernet uses the Internet Protocol (IP), and for Layer 4 it uses the Transmission Control Protocol (TCP). Hence, a distinguishing feature of Ethernet (in comparison with other fieldbuses) is that it uses the TCP/IP protocol.

One of the standard configurations for Ethernet is 10BASE-T, which provides a 10 megabit per second (Mbps) rate running over unshielded twisted-pair wire in a star topology. The hubs in the star topology can be either repeater hubs or switching hubs. If switching hubs are used, the network is referred as *Switched Ethernet*. An advantage of Switched Ethernet is that if bits can be transmitted simultaneously in both directions on all the links, then there are no collisions of messages since the paths from port-to-port in a switch are independent. However, switches are more expensive than repeaters, and thus repeater networks are more likely to be employed as fieldbuses.

Overall Network Configuration

In controlling a large collection of devices and equipment as in a factory, the standard structure of the overall communications network is to have three levels consisting of the sensor/device network level, the control network level, and the plant or factory network level. An illustration of the overall network configuration is shown in Figure 11.5. The top level of the network is sometimes referred to as the *backbone* of the network.

As noted earlier, Ethernet or Switched Ethernet 10BASE-T can used use for the control network. To achieve greater bandwidth capability, especially for the plant network level, *Fast Ethernet* can be used in place of 10BASE-T. Fast Ethernet is based on the 100BASE-T standard that provides a 100-Mpbs rate over twisted pair or optical fiber. Before the year 2000, Gigabit Ethernet will be available which will provide a 1- Gbps rate, and thus should meet the requirements for the backbone in most applications.

In addition to Ethernet, other LAN technologies can be used in control/manufacturing environments. These include Fiber Distributed Data Interface (FDDI), 100VG-AnyLAN, and Asynchronous Transfer Mode (ATM). With the upgrades that are continuously being generated for Ethernet (such as Gigabit Ethernet), it appears that Ethernet will continue to be a strong contender as the LAN of choice in many application domains.

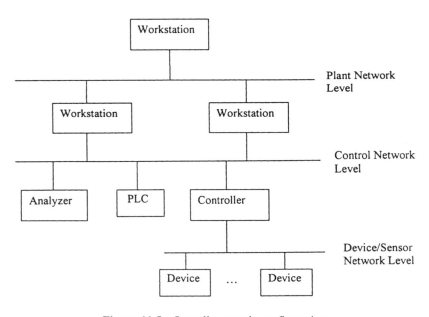

Figure 11.5   Overall network configuration.

Web-Based Studies

For the following tasks, generate a report using information acquired from the web. Your report should be typed using word processing software and be limited to a maximum of five pages.

11.1   Give a more detailed explanation of the OPC standard for equipment interfacing than is given in this chapter.

11.2   Name the advantages and disadvantages of the four different types of MAC for sensor and device networks used in control.

11.3   Give a detailed description of 10BASE-T.

11.4   Compare Ethernet and Fast Ethernet.

11.5   Describe Switched Ethernet and discuss its applicability to control networks.

11.6   Describe those network performance parameters that are particularly important in the application to control.

Appendix A

Further Reading

For additional information on the topics covered in this book, the reader may consult the references given below.

Continuous-Variable Control

Dorf, C. R., and Bishop, R. H., *Modern Control Systems*, 8th ed. Addison Wesley, Reading, MA, 1997.

Phillips, C. L., and Harbor, R. D., *Feedback Control Systems*, 3rd ed. Prentice Hall, Englewood Cliffs, New Jersey, 1995.

Franklin, G. F., Powell, J. D., and Workman, M. L., *Digital Control of Dynamic Systems*, 2nd ed. Addison Wesley, Reading, MA, 1990.

Phillips, C. L., and Nagle, T. H., *Digital Control Systems Analysis and Design*, 3rd ed. Prentice Hall, Englewood Cliffs, NJ, 1994.

Franklin, G. F., Powell, J. D., and Emami-Naeini, A. R., *Feedback Control of Dynamic Systems*, 3rd ed. Addison Wesley, Reading, MA, 1990.

Advanced Control Techniques

Camacho, E. F., and Bordons, C., *Model Predictive Control*, Springer-Verlag, NY, 1998.

Mosca, E., *Optimal, Predictive and Adaptive Control*, Prentice Hall, Englewood Cliffs, New Jersey, 1995.

Omidvar, O., and Elliott, D. L., *Neural Systems for Control*, Academic Press, NY, 1997.

Seborg, D. E., *Nonlinear Process Control*, Prentice Hall, Englewood Cliffs, NJ, 1996.

Discrete Logic Control

Bryan, L. A., and Bryan, E. A., *Programmable Controllers: Theory and Impl*ementation, 2nd ed. Industrial Text Co., Atlanta, GA, 1997.

Hughes, T.A., *Programmable Controllers*, ISA, NY, 1997.

Parr, E. A., *Programmable Controllers*, 2nd ed. Butterworth-Heinemann, Oxford, 1998.

Webb, J. W., Programmable Controllers: Principles and Applications, MacMillan, NY, 1990.

Manufacturing Systems and Production Control

Elsayed, A., and Bouchar, T. O., *Analysis and Control of Production Systems*, 2nd ed. Prentice Hall, Englewood Cliffs, NJ, 1993

Groover, M. P., *Automation, Production Systems, and Computer Integrated Manufacturing*, 2nd ed. Prentice Hall, Englewood Cliffs, NJ, 1990.

Hopp, W. J., and Spearman, M. L., *Factory Physics*, Prentice Hall, Englewood Cliffs, NJ, 1996.

Mahoney, R. M., *High-Mix Low-Volume Manufacturing*, Prentice Hall, Upper Saddle River, New Jersey, 1997.

Computer Communication Networks

Breyer, R., and Riley S., *Switched and Fast Ethernet*, 2nd ed., Ziff-Davis Press, Emeryville, CA, 1996.

Halsall, F., *Data Communication, Computer Networks and Open Systems*, 4th ed. Addison Wesley, Reading, MA, 1996.

Tannenbaum, H. S., *Computer Networks*, 3rd ed., Prentice Hall, Englewood Cliffs, NJ, 1996.

Appendix B

Laboratory Project

For students taking a course based on the material in this book, it is highly desirable to require them to do a laboratory project on the design and implementation of both continuous-variable and discrete logic controllers. In carrying out the project, students obtain hands-on experience in using the "tools of the trade" in industrial controls such as PLCs, real-time PC-based controllers, and software for equipment interfacing, operation, and communications. To provide a physical platform for the project, a process demonstrator consisting of plastic tanks connected by plastic pipes with pumps and valves was designed and built. The project deals with the movement and "processing" of fluid in the tank system. Details for a two-tank system and experiments based on the system are given in this appendix. The cost of building the system described here depends on the quality of the components and the amount of the educational discount obtained from vendors. With high-quality state-of-the-art components and a good discount, the cost should be less than $20,000.

Process Demonstrator

The process demonstrator is a "scaled closed version" of a fluid processing system typically found in industry such as in a chemical production plant. At Georgia Tech, a three-tank system was built; however, a two-tank system is sufficient for performing experiments involving continuous-variable and discrete logic control. The process and instrumentation diagram (P&ID) of a two-tank system is shown in Figure B.1. Each of the tanks can hold 25 gal. Fluid enters and leaves each tank through pipes located at the bottom of the tank (the P&ID shows fluid entering tanks at the top). The output line of each tank has a Goulds pump, denoted by P 103 and P 204 on the P&ID. There are flow valves (FV 202 and FV 205), orfice plates (FE 202 and FE 205), and flow meters (FT 202 and FT 205) on the input and output lines of Tank 2. The flow valves (Fisher globe valves with Jordan Controls electric rotary actuators) are proportional valves with a linear range of operation from 0% (closed) to 100% (open). A solenoid on/off valve (SV 104) is on the line coming out of Tank 1 and there is a manual valve on a second line into Tank 2. Tank 2 has a mixer for stirring the fluid in the tank, and there are four heating elements located at the bottom of the tank.

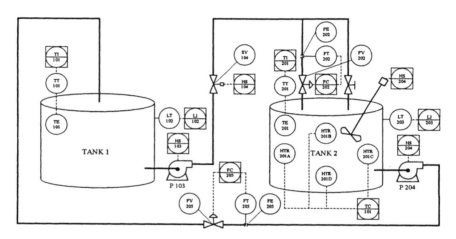

Figure B.1 Process and instrumentation diagram of two-tank system.

The level of fluid in each tank and the flow rate into and out of Tank 2 are measured using differential pressure transmitters (LT and FT). The temperature in each tank is measured using resistance temperature detector elements (TE) and temperature transmitters (TT). Signals from the two-tank system consist of readings of fluid flow rate, fluid level, and fluid temperature. The signals are transmitted using Honeywell Smart Transmitters that operate in either the analog 4-20 mA format or in a digital format (STIM). Signals sent to the two-tank system consist of analog commands to the flow valves and digital commands to turn on or off the pumps, mixer, heaters, and the solenoid valve in the output line of Tank 1. The instrumentation in the two-tank system is interfaced to a PC using National Instruments DAQ hardware (a SCXI-1000 Signal Conditioning Module, A/D and D/A boards) and LabView/Lab Windows CVI software packages. The instrumentation is also interfaced to a Honeywell industrial controller containing a PLC. The continuous-variable control part of the project uses the PC interface, while the discrete logic control part uses the commercial industrial controller. As a result, students obtain exposure to both PC-based and PLC-based interfaces.

Interlocks must be programmed into the two-tank system to prevent the possibility of overflow or underflow of fluid in the tanks. A minimum level of fluid must be maintained in the tanks to prevent the pumps from drawing air. Interlocks are also necessary to prevent fluid overheating and dead heading of the pumps (i.e., pumps are on only when the valves are at least 10% open). For the PC interface, interlocks are coded in LabView and designed into a LabView template that is provided to the students. Output signals first pass through the interlock code before energizing the actual hardware. Interlocks for the Honeywell industrial controller are designed using ladder logic and continuous loop operations.

Project Description

The students work on the project in teams of three or four. The content of the project depends on the amount of time to be devoted to the project. In the span of one semester with a one-credit-hour laboratory project, students should be able to complete all three of the modules mentioned here. For a one-quarter format, students should be able to complete the level-control module and the recipe module. Of course, the modules described here can be modified or other modules can be added. By scheduling the student teams at different times for use of the laboratory, it is possible to accommodate up to eight teams (32 students) with a single two-tank system (five hr per week per team over a 40-hr week).

For carrying out the following tasks, the total amount of fluid in the system should be equal to the amount of fluid required to fill one tank plus the minimum amount of fluid required in the second tank to prevent the pump in the tank's output line from drawing air.

Level Control

With pump P 204 off and valve FV 205 closed in the output line of Tank 2, use a waveform chart to record the step response for the filling of Tank 2. Using the approach in Chapter 3, determine a first-order continuous-time model with time delay for the filling of Tank 2. Using the techniques in Chapter 5, discretize the tank model and then design a modified digital PI controller that gives the fastest possible response in reaching a desired level without exceeding the limit on the input valve position (maximum open position is 100%). Implement the controller design using LabView and evaluate the performance of the controller by opening output valve FV 205 and turning on pump P 204 in the output line of Tank 2. If necessary, modify the controller gains until an acceptable performance is achieved.

Temperature and Level Control

With pump P 204 off and valve FV 205 closed in the output line of Tank 2, and with the fluid level in Tank 2 equal to some desired value, use a waveform chart to record the step response for heating the fluid in Tank 2. Using the approach in Chapter 3, determine a first-order continuous-time model with time delay for the process of heating the fluid in Tank 2. Using the techniques in Chapter 5, discretize the tank model and then design a modified digital PI controller that gives the fastest possible response in reaching a desired temperature level without exceeding the limit on the duty cycle of the heating elements (maximum duty cycle is 100%). Implement the controller design using LabView and evaluate the performance of the controller by opening output valve FV 205 and turning on pump P 204 in the output line of Tank 2. Use a level controller to keep the level in Tank 2 at the desired value. If necessary, modify the controller gains until an acceptable performance is achieved. Determine how sensitive the temperature controller is to the volume of fluid in Tank 2.

Recipe Control

Design a discrete logic controller that carries out the following sequence of operations:

1. Fill Tank 2 to 75% of full capacity.
2. Turn the mixer on for 30 sec.
3. Move 50% of the fluid in Tank 2 to Tank 1.
4. Hold the fluid in Tank 1 for 30 sec.
5. Refill Tank 2 to 75% of full capacity.

Using the approach in Chapter 7, define the inputs, outputs, and state variables for the controller, and for each state variable, determine the state-transition diagram. From the state diagrams, determine Boolean logic equations for the desired control action, and from this, develop a ladder logic program using the results in Chapter 8. Implement the controller using the PLC in the industrial controller and evaluate the performance.

Index

A
A/D conversion, 76
Adaptive control, 102-106
Automation, 3

B
Balance delay, 183-184
Balancing, 181-186
Bandwidth, 75-76
Blocking, 179, 200
Boolean algebra, 121-123
Boolean logic equations, 121, 123-134, 153-154, 160-161
Bottle-filling operation, 156-162
Bottleneck machine, 176, 178

C
Capacity, 170-171
Circuit board assembly, 1-2
Client/server architecture, 205-206
Closed-loop control, 5, 11-12, 13-15, 16-22, 47-69, 75-76, 80-90
Closed-loop transfer function
    continuous-time, 17, 48-50
    discrete-time, 85-87
Communication networks, 4, 209-214
Computer integrated manufacturing (CIM), 3
Continuous-flow manufacturing, 2
Continuous variables, 5-6
Continuous-variable control, 12-15, 16-22
Control
    adaptive, 102-106

continuous variable, 12-15, 16-22
dead-beat, 88-90
digital, 75-76, 80-90
digital PI, 80-87
discrete event, 10-12
discrete logic, 12, 113-121, 128-134
introduction to, 3-6
inventory, 201-203
model predictive, 95-102
modified PI controller, 53-54
modified PID, 72-73
neural net, 106-110
optimal, 99-100
PD, 71-72
PI, 51-59
PID, 59-66
production, 193-203
proportional, 14-15, 16-22, 51-52, 66-69
recipe, 10
sequencing, 10, 114
Control applications
bottle-filling operation, 156-162
dc motor, 30, 56-62, 64-66, 71-73, 83-90
gas furnace, 7, 25, 71
level control, 10-15, 17-22, 66-69, 114-117, 130-131, 151, 221-222
oven, 7
Control relay, 143-147
CONWIP, 196-200
Cycle time, 178, 182-183

D
D/A conversion, 76
Damping ratio, 36-38
dc motor, 30, 56-62, 64-66, 71-73, 83-90
Dead-beat control, 88-90
Derivative control, 60-62
Design for manufacturing (DFM), 2
Difference equation, 78-79
Digital control, 75-76, 80-90
Direct adaptive control, 104-106
Discrete event control, 10-12
Discrete event system, 6

Discrete logic control, 12, 113-121, 128-134
Discrete parts manufacturing, 1-2
Discrete-time system, 78-79
Discrete variables, 5
Distributed control system (DCS), 114
Disturbance
  input, 62-66, 98
  estimator, 98-99

E
Equipment interfacing, 205-209
Estimator
  disturbance, 98-99
  system parameters, 102-104
  controller parameters, 105-106
Ethernet, 213-214
Event-driven system, 6,11, 123
Excitation condition, 104, 106

F
Fieldbus, 212-213
Field devices, 113-114
Finite-dimensional process, 30-31
First-in first-out (FIFO) policy, 173
Flow line, 165-166, 171-186
Flow shop, 165-166
Full-on/full-off production, 194-195

G
Gas furnace, 7, 25, 71
GEM standard, 207-209

H
Hedging point, 202
Hold operation, 75, 83-84
Hybrid system, 6

I
Indirect adaptive control, 102-104
Interfacing, 4, 205-209
Inventory, 194
Inventory control, 201-203

Iteration, 79

J
Job shop, 165-166

K
Kanban, 200-201

L
Ladder logic diagram
  electrical, 141-147
  software, 147-151, 153-155
Laplace transform, 15-16
Latching, 130, 146-147, 149
Level control; See Control applications
Linear operation, 27-28, 78
Line efficiency, 178-181
Little's law, 177, 197

M
Manufacturing control hierarchy, 4, 165
Manufacturing delay, 179
Manufacturing examples
  circuit board assembly, 1-2
  metal sheeting production, 2-3
Manufacturing fundamentals, 1-3, 165-166
Manufacturing lead time (MLT), 167-168, 195-196
Material requirements planning (MRP), 195-196
Medium access control (MAC), 212-213
Memoryless, 124
Metal sheeting production, 2-3
Model error, 98
Modeling, 27-41
Model predictive control (MPC), 95-102
Modified PI controller, 53-54

N
Natural frequency, 36-38
Network protocols, 211-212
Neural net controllers, 106-110
Neuron, 108

O

Optimal control, 99-100
Order, 31, 79
Oven, 7
Overshoot, 35-38

P

Pade approximation, 39-41, 67-69
Peak time, 36-38
PI controller, 51-59
PID controller, 59-66
Poles, 17, 31, 49-50, 87
Production control, 193-203
Production performance measures, 166-171
Production rate, 168-170, 173, 178, 194-195
Programmable logic controller (PLC), 141-142, 151-153, 161-162
Proportional controller, 14-15, 16-22, 51-52, 66-69
Pull system, 196-201
Push system, 196
Push-and-pull system, 201-203

R

Recipe control, 10
Reference signal, 13, 50, 63
Relay
    control, 143-147
    software, 149, 157
Release times, 193
Reset/set operation, 124-132, 149, 154-155, 158, 160-161

S

Sampled-data controllers; see Digital control
Sampling frequency, 75-76
Sampling interval, 75, 77
Sequencing control, 10, 114
Setpoint, 50
Settling time, 36-38
Software relay, 149, 157
Stability
    process control, 50, 87
    production, 195
Starvation, 179

State diagrams, 114-121, 128-130, 132-133, 157-159, 207-208
State variable representation, 123-134
Step response, 31-38, 55

T

Tank system, 9-10, 219-223
Throughput, 173
Time constant, 19-21, 32, 58
Time delay, 12-13, 38-41, 66-69
Time driven system, 6
Time invariance, 28, 78
Timer coil, 157-158
Tracking, 47, 50-66, 87-90, 96
Transfer function representation
  continuous time, 28-31
  discrete time, 78-79
Transfer line; see Flow line
Transient response, 19-22, 50-51, 55-56

U

Underdamped, 36-38
Unit disk, 87-88
Unity feedback, 14, 47

W

Weighting factor, 99
Work content, 182
Work-in-process (WIP), 171, 175-178, 197-198

Z

Zeros, 31, 49-50
z transform, 77